PRIMATES OF THE WORLD

Rod and Ken Preston-Mafham

BLANDFORD

A BLANDFORD BOOK

This paperback edition first published in the UK 1999 by Blandford
A Cassell imprint

Cassell plc,
Wellington House
125 Strand
London WC2R 0BB

www.cassell.co.uk

Hardback edition published in the UK by Blandford 1992

Distributed in the United States by Sterling Publishing Co., Inc.,
387 Park Avenue South, New York, NY 10016-8810

A Cataloguing-in-Publication Data entry for this title is available and may be
obtained from the British Library

ISBN 0-7137-2791-8

Typeset by August Filmsetting, Haydock, UK

Printed in Hong Kong by Colourcraft Ltd

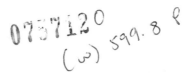

PRIMATES
OF THE
WORLD

Contents

Acknowledgements

A number of people have earned our thanks for rendering assistance during Ken's tropical travels. In Madagascar Patrick Daniels of Duke University, N. Carolina, helped in a number of ways at Ranomafana, where Ken was also privileged to accompany Deborah Overdorff into the forest to observe and photograph her habituated groups of red-fronted and red-bellied lemurs. On Nosy Mangabe Eleanor Sterling gave freely of her time and knowledge to enable him to photograph the aye-aye in the wild.

When writing a book of this kind it is inevitable that the authors have made extensive use of information from colleagues who have made their work available to the scientific fraternity via papers published in specialist journals. It is not possible to list all the references individually here, but the names of all the authors are as follows: Adamson; Agnagna; Anadu; Andau; Anderson; Antinucci; Bartecki; Bielert; Boase; Boesch; Boinski; Bourlière; Brewer; Brown; Byrne; Chapman; Constable; Cords; Crompton; Deng; Denning; Dewar; Dutrillaux; Fairbanks; Fay; Fedigan; Fernandez; Ferrari; Galdikas; Gautier-Hion; Girolami; Glander; Goldizen; Goldstein; Haimoff; Harcourt; Henzi; Heymann; Hiraiwa-Hasegawa; Howarth; Hutchings; Izard; Kappeler; Kaufman; Kinzey; Koenders; Koman; Macedonia; McGrew; Mertl-Millhollen; Mittermeier; Moore; Morland; Nash; Newton; Nishida; Norconk; Oates; O'Brien; Oko; Overdorff; Pereira; Peyrieras; Podolsky; Pollock; Rajpurohit; Ramangason; Randrianasolo; Ratsirarson; Richard; Robinson; Rowell; Ruempler; Sauther; Schurmann; Seeligson; Seigler; Simons; Small; Smith; Soini; Sommer; Sow; Srikosamatara; Stanford; Stewart; Sugardjito *et al*; Sugiyama; Sussman; Takahata; Takasaki; Tattersall; Thierry; Tilson; Toyama; Tutin; Valderrama; van Hoof; van Noordwijk; van Schaik; Visalberghi; Wahome; Warter; Watts; Whitten; Wilson; Wright; Wunderlich; Yeager; Zhao; Zunino.

We would also like to thank the staff of Twycross Zoo, where we took our photographs of captive primates and were provided with useful information; and the Monkey Sanctuary in Cornwall, who gave us information on the rehabilitation of woolly monkeys into the wild. We are also indebted to Chris O'Toole of the University Museum in Oxford for obtaining many of the necessary scientific journals for us.

Finally, we must thank Rod's wife Jean, who has undertaken the onerous task of preparing the index.

Preface

Much has already been written about primate biology and behaviour, and several excellent books on the subject have appeared in recent years. However, it is in the 1970s and 1980s that the study of primates in their natural environment has really taken off. This has resulted in a deluge of new and often fascinating information on the lives of primates in the forests and mountains of the tropical world, rather than on those kept in the cage or enclosures of zoos and breeding projects. In writing the present book we have taken advantage of as much of this newly published material as possible, so as not to overlap too closely with earlier published work. As a result we have had to be quite selective in our handling of subjects, treating some in great detail where they are interesting and recent information is available, rather than trying to cover lots of different aspects superficially. On the other hand, because we are particularly interested in the life of primates in the wild we have kept to a minimum any mention of primate learning which has resulted from changes in the environment introduced by man (such as 'washing' behaviour in Japanese macaques).

Ken has spent many years travelling in the tropics and is very familiar with the joys and drawbacks of studying and photographing primates in the wild. Close-up photography of many monkeys and lemurs in their tropical forest homes is now becoming easier and easier; this is due almost entirely to the dedicated efforts of academic workers who have gone alone into an area and habituated the study group of animals to close human presence. Sometimes, as in the case of certain lemurs which have never been hunted by man due to local taboos, the process of habituation may take only a few weeks – or perhaps even days. On other occasions the process may be heartbreakingly prolonged, difficult and laborious. Attempts to get close to animals which are naturally nervous and have a long history of being hunted, such as the pig-tailed macaque in the Malaysian forests, provide a good example. The achievements of these researchers, often only in their early twenties, should not be underestimated. They have both contributed to our knowledge of primates – much of this book has been based on their published work – and enabled visitors to experience the thrill of approaching to within a metre or two of a group of wild primates.

Even when the study group is well habituated, life can still be difficult. Ken remembers an episode in a Madagascan forest when a group of diademed sifakas had raced down through the tree canopy on a very steep slope – only to come hurtling back up again a few minutes later, just as the unfortunate humans beneath had finally managed to negotiate the slippery slopes, tangled tree-falls and snagging lianas to reach them at the bottom. As the group of animals disappeared upwards again one of the American researchers, out of breath and covered in sweat, summed up his frustrations with the words: 'I hate these damned animals and I hate this forest.' Of course, he only meant it at that moment; and a few minutes later, after slogging back up that same slope, both of us were delighting in the experience of standing next to the sifakas as they gambolled on the ground, now and again coming right up to us and peering up as if inviting us to join in their game. No doubt those months in the forest will, in later life, become one of the most treasured experiences for that momentarily dispirited lemur-watcher.

Primates of the World makes one clear departure from our previous books in this series, which have covered various forms of invertebrate animals. With patience and the necessary expertise, small creatures such as spiders, butterflies and grasshoppers can usually be approached very closely in the wild, and photographed to a very high standard. Primates are different. As mentioned above, considerable effort has to be made to allow a close enough approach for study, let alone for high-quality colour photography. Many species have simply never been habituated or, even after years of study, will still not allow anyone to get close enough, in their dense, dark forest environment, to take high-quality pictures. It would not be sensible to omit large numbers of important groups or species merely because it has not been possible to obtain good photographs of them in the wild, and so we have taken what is for us the unusual step of featuring some of them in captivity. This is particularly important where details of the face are needed – in the wild, this can only be achieved with the tamest of primates.

To assist those readers who are not too familiar with the various primates, when we introduce a species for the first time we give it its common name followed by its scientific name – for example, the indri (*Indri indri*) – but from then on we only use common names. If at any time you need to cross-reference these names, refer to the index.

Ken Preston-Mafham
Rod Preston-Mafham
King's Coughton, England

Chapter 1

Introduction to the Primates

Ask the average person what they understand by a primate, and they are likely to reply that it is a monkey or an ape. Both of these answers are correct, but the order Primates also includes, or once included, a number of other groups of animals as well. The present-day understanding of the order is that it encompasses six natural groups – the lemurs; the lorises and galagos; the tarsiers; the New World monkeys; the Old World monkeys; and finally apes and humans. The first three of these groups are often collectively referred to as the prosimians while the remainder are the simians, although in reality the tarsiers fit uncomfortably within the simians and are thought of by many people as an intermediate between the prosimians and simians. At various times the tree shrews have also been included in the Primates, but current trends are to keep them separate in an order of their own. For a more detailed discussion of the various members of the order Primates, see p. 26.

When placing mammals into a particular order, taxonomists look for particular evolutionary specializations and adaptations which are characteristic of the group. The artiodactyls, for example, which include sheep, goats, cattle and deer, have cloven hooves as a distinguishing characteristic. What, therefore, differentiates the order Primates from the other mammalian orders? Structurally speaking, the answer is: 'Not a lot'; behaviourally speaking, however, the answer is: 'A great deal.'

Structurally, primates show very little deviation from the basic primitive mammalian pattern: they retain a collar bone or clavicle and their hands and feet still have five digits, with, in a few species, a limited reduction of the bones within each digit. One very obvious primate feature is the development of the forward-facing eyes at the expense of a reduced muzzle and an accompanying reduction in the sense of smell. At the same time, the movement of the eyes on to the front of the head has allowed expansion of the braincase and the enlargement of the brain. It is this combination of a highly mobile skeleton of primitive design but with a greatly developed brain which has led to the success of the primates as a group.

Having briefly mentioned the basic structural characteristics of the primates, we will now look at them in more detail.

SKELETAL FEATURES

The primate skeleton maintains many of the features of the early mammals – features which are advantageous to a group of animals which are in the main arboreal. Notable are the continued presence of the clavicle or collar bone, a

The South American squirrel monkey (*Saimiri sciureus*), the traditional organ-grinder's assistant, probably represents to most people the archetypal primate. Photographed at Twycross Zoo.

separate radius and ulna in the front limbs and separate tibia and fibula in the hind limbs, although in the tarsiers the latter bones are partially fused. Keeping the clavicle confers a distinct advantage, for it enables primates to hang from tree branches without putting a great deal of strain on their shoulder muscles. The retaining of a separate radius and ulna permits pronation and supination – rotation through 180 degrees – of the forearm, i.e., from palm down to palm up. This is very important to animals which use their hands to find and hold food and to manipulate objects within their environment, and for those primates which move through the trees by swinging from branch to branch – known as brachiation.

In relation to the handling of objects, the hands and feet show very little reduction from the bone count of primitive mammals; as a consequence long fingers and toes are found in most primates, although there has been a loss of the thumb in a few monkeys and some Lorisidae. To a greater or lesser extent, depending upon the family to which they belong, primates have an opposable pollex (thumb) and/or an opposable hallux (big toe). True opposability of the thumb reaches its ultimate in man, whose thumb can rotate about its axis to the point where its pad completely opposes that of the index finger. This is also found in the families Hylobatidae, Cercopithecidae and Pongidae. In the remaining families the thumb is pseudo-opposable – that is, it is able to move away from the other fingers in the same plane to a fairly wide angle. The thumb can then be moved back towards the fingers, giving some degree of grip of any object lying between it and the fingers. All primates apart from man also have opposable big toes, a feature which enables them to grasp the branches upon which they stand. One interesting adaptation in the tarsiers and galagos is the elongated tarsal region in the hind limb. This increases the length of the foot, giving greater leverage for leaping, but without impeding the animal's ability to grip branches with its toes – which would be the case if the finger bones were elongated, as they are in other leapers such as kangaroos.

The vertebral column is of the basic mammalian design, with the main differences between primate families being reflected in the degree of development of the tail. While the tail is reduced or absent in the great apes, man and some monkeys, it is well developed in the remainder of the primates. The prosimians and many monkeys have long tails, though truly prehensile tails are found in just two subfamilies of New World monkeys, the Atelinae and the Alouattinae. In a prehensile tail up to one-third of the lower surface at the tip of the tail consists of bare skin, and this gives the monkey a very strong grip when the tail is curled round a branch. As a result, the spider, woolly and howler monkeys are able to hang from branches suspended by the prehensile tail alone. Other primates can also wrap their tail around a branch whilst sitting at rest, but this is only to help them maintain their balance.

A special adaptation of a small part of the vertebral column is found in the African potto (*Perodicticus potto*). The dorsal spines on the lower neck and first thoracic vertebrae are very elongated, and along with the muscles in this region they give the potto extra protection if seized by a potential predator.

Although, with the exception of man, primates are quadrupedal, they do

All primates, apart from man, have opposable big toes. This feature is clearly visible on this female red howler monkey (*Alouatta seniculus sara*) photographed at Twycross Zoo.

The tip of the prehensile tail of a black spider monkey (*Ateles paniscus*) showing the bare, 'fingerprinted' skin which is used to grip tightly round branches.

show a considerable degree of variation in limb length; this is related to the way in which they move through their arboreal habitat. Those primates which clamber around the trees in a basic quadrupedal fashion are usually relatively short-limbed. Ground-living primates, who may need to run to the safety of trees or rocks to escape their enemies, tend to be fairly long-limbed. Highly arboreal primates such as the gibbons, which swing along branches and from branch to branch, have long fore-limbs; while those which leap from branch to branch and from tree to tree and cling vertically to them tend to have elongated hind-limbs.

The structure of the skull shows the primates to be a group which depends more upon sight than smell as a means of obtaining information from the outside world. The eyes are enclosed in bony sockets and are generally on the front of the face, in a position which in many primates has allowed the evolution of overlapping visual fields and consequent binocular vision. As a result the muzzle tends to have become shortened, reducing the volume available for the sense of smell, though in the lower primates, especially the lemurs, the eyes are not completely frontal and the muzzle is still quite prominent. A noticeable difference between the skulls of the platyrrhines and the catarrhines (see p. 28) may be seen in the region of the ear. In the latter there is a bony tube, the external auditory meatus, leading to the middle ear; in the platyrrhines this tube is absent.

True bipedal gait in primates is restricted to humans, but when moving across open areas between trees Verreaux's sifaka (*Propithecus verreauxi verreauxi*) does its best by performing a kind of bouncing, dancing action in which the hands and arms are flung around as balancing organs. Photographed in Madagascar.

TEETH

Many primates have omnivorous diets, often with the emphasis on plant food, though a few are purely herbivorous and the tarsiers eat only animal food. As a result primate dentition shows very little specialization of the type found in other groups of mammals, and all primates retain at least one of the four main tooth types – incisors, canines, premolars and molars – in each side of the upper and lower jaws. No modern primates retain the total of forty teeth to be found in the jaws of primitive primates such as the Eocene lemur *Notharctus*. The prosimians and New World monkeys come closest, with a total of thirty-six teeth, a premolar having been lost from both upper and lower jaws on each side. Marmosets have only thirty-two teeth – a molar has been lost on each side of both upper and lower jaws; catarrhines have the same number of teeth, except that they have lost a premolar instead from each side of both jaws. This gradual reduction in the total number of teeth reflects the gradual reduction in the length of the primate muzzle, since shorter jaws can obviously accommodate fewer teeth. In the baboons, with their elongated muzzle, there has been an accompanying elongation of the molars to occupy the extra space.

A few individual species have lost further teeth: for example, the aye-aye has a total of only eighteen. Its incisors have been reduced to a single pair in the upper and lower jaws; those in the lower jaw are similar to those of rodents in that they are long, grow continuously as they wear away, and are used for gnawing. Canines are absent from both jaws, and a single premolar is maintained on each side of the upper jaw alone; it is only the molars that remain unchanged, with three on each side of both upper and lower jaws.

An interesting modification of the lower incisors and canines is to be found

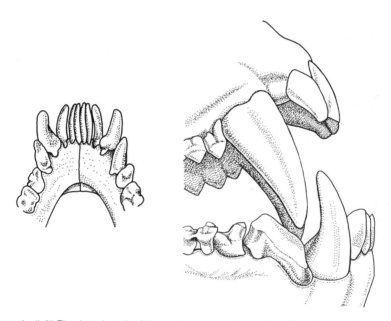

Figure 1 (left) The dental comb of the prosimians, which consists of the flattened, forward projecting incisors and canines of the lower jaw, is used for grooming and cleaning the fur.

Figure 2 (right) In a number of species of Old World monkeys the males have exceptionally well-developed upper canine teeth, which are honed by the flattened occlusal surface of the first premolar tooth on each side of the lower jaw. In addition they have a gap, the diastema, between the upper canine and the adjacent incisor; the lower canine fits into it when the jaws are closed.

in the prosimians (with the exception of the aye-aye). In the prosimians these teeth are elongated, compressed laterally and point forwards from the lower jaw almost horizontally to form a dental comb which is used in grooming the fur. In addition, the first premolar is caniniform, that is, elongated to a sharp point, to replace the canine proper which is in the dental comb. Primates with a dental comb also possess a sublingual organ, a horny plate on which there are tiny teeth; it lies below the tongue and is used to keep the dental comb free of accumulated bits of hair and other debris which becomes lodged there during grooming.

With the exception of humans the canine teeth in primates are well developed, so that they extend above the tops of the other teeth in both jaws. As a consequence there are gaps in the upper and lower jaws to accept the points of the opposing canine teeth. In anthropoids things are carried a step further, for the first premolar in the lower jaw is furnished with a sloping facet which hones the large upper canine as the jaws close. The drawback to these large canines is that they prevent the side-to-side chewing of food, which is typical of humans with their stunted canines.

Although there are minor differences in the origins of the cusps on the primate cheek teeth, they are basically uniform in nature – a reflection of their mainly omnivorous diet. A very noticeable exception is in the gelada (*Theropithecus gelada*), which is completely herbivorous. Its cheek teeth have well-defined ridges which are used to mince up the grass which is its main food.

EXTERNAL FEATURES

All primates are covered to a greater – or, in the case of humans, lesser – extent in hair, a uniquely mammalian structure. The nature of this hair varies within the primates; it may be woolly, silky or, as in the great apes, quite coarse. The primary function of the variations in colour and patterning of this hair is to enable the primates to recognize their own and other species. At the same time it enables us to distinguish the many species and subspecies. Where there is sexual dichromatism, the different markings on the males and females allow them to recognize the opposite sex. In a number of primates the hair on the head may be extended in the form of tufts or may form distinctive crests, beards or even moustaches, while in some species the body hair is elongated to form cloaks or manes.

Not all of the skin is covered in hair in all primates: a number have bare patches on the face and/or around the genitals. The face of the New World red uakari (*Cacajao rubicundus*), for example, is almost hairless and is bright red in colour, while a number of Old World genera also have naked faces. The bare, often brightly coloured skin around the genital region of a number of genera of Old World monkeys is part of their sexual display.

These patches of coloured skin on the rear end should not be confused with the ischial callosities to be found among the cercopithecid monkeys and the gibbons. These are special areas of cornified skin devoid of hair and are actually fused to the ischial bones of the pelvic girdle. These particular

The characteristically well-developed canines of many male Old World primates are starkly revealed in this shot of a toque macaque (*Macaca sinica*) yawning as he rests in a Sri Lankan forest.

The naked face of the red uakari (*Cacajao rubicundus*), its long, shaggy hair, its short tail and its odd posture when walking combine to produce a rather bizarre-looking monkey. It is a specialist nutcracker capable of piercing some remarkably tough shells. Photographed at Twycross Zoo.

primates spend a lot of time sitting, and the callosities provide them with a comfortable 'seat'.

An external feature which is very typical of primates is the presence of flattened nails rather than claws on the ends of the digits. There are, as usual, a few exceptions to this general rule. The New World marmoset monkeys have, for example, claws rather than nails, and the aye-aye has claws on all digits except the big toes which bear nails instead. In order to help in scratching through their fur all the prosimians have a so-called toilet claw on the second digit of the foot, while tarsiers have one on both the second and third digits of the foot.

PRIMATE PHYSIOLOGY

There is little that needs to be said about the internal workings of primates, such as the circulation, respiratory system and so on, for they are in principle the same as those of other mammals. The alimentary canal is very basic in design, showing little of the modifications that are found, for example, in the ruminants. In some of the highly herbivorous primates which feed mainly on leaves there is, however, some enlargement of the stomach to cope with the large bulk of food taken in, and the caecum is well developed in many primates. At least one species, the sportive or weasel lemur (*Lepilemur mustelinus*), is said to be coprophagous – that is, it reingests special droppings produced from the caecal contents, which are then passed through the gut for a second time to be digested. In this way it is able to make efficient use of the highly indigestible leaves upon which it feeds; bacteria in the caecum carry out the necessary digestion before reingestion takes place.

A feature of one monkey subfamily, the Cercopithecinae, is the presence of pronounced cheek pouches which allow the monkeys to feed rapidly and then to chew and swallow their food at their leisure. This ability is clearly of advantage to those members of the subfamily which are mainly terrestrial, for while they are feeding they are less aware of any approaching predator.

The female reproductive organs of the higher primates show a marked difference from the normal mammalian pattern, and this is related to the fact that in most instances they give birth to a single offspring. In the majority of mammals the uterus has two horns, but in the most advanced primates it has a single chamber and is called unicornuate or simplex. The bicornuate uterus is still to be found in the prosimians, but even within this group there is a range of intermediate types leading towards the simplex form.

The structure of the primate penis also shows some marked variations. With the exception of humans and two platyrrhine genera a small penis bone or *os penis*, typical of most mammals, is retained by the primates. In some of the prosimians there is considerable variation in the morphology of the tip of the

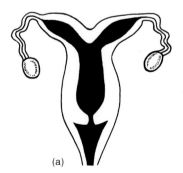 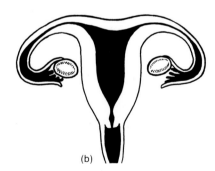

(a)   (b)

Figure 3 Prosimians and tarsiers in general possess the more primitive bicornuate uterus (a), while the simians tend to have the more advanced unicornuate or simplex uterus (b). Within the primates, however, may be found transitional forms between these two extremes.

penis – so much so that they can be used as a means of identifying individual species.

The most advanced form of placentation is found in the anthropoidea and tarsiers. The efficiency of a placenta is governed by the number of cell layers through which oxygen and food in one direction, and waste products in the opposite direction, have to pass, when diffusing between the mother's blood and that of her offspring. In the most primitive placentas these products have to pass through the walls of both the placental and the uterine blood vessels as well as the placental and uterine epithelia. As the mammals have advanced the number of these barriers has gradually reduced, until in the anthropoids and tarsiers we find the haemochorial placenta. Here, the placental blood capillaries burrow into the wall of the uterus where they come into direct contact with the mother's blood. As a result the only separation between the mother's circulation and that of her offspring is the wall of the placental capillary. The prosimians have a less efficient placenta, with the blood of the mother and offspring separated by the walls of the placental capillaries, the uterine epithelium and the walls of the maternal capillaries – three layers of cells, as against a single layer in the more advanced placenta.

The primate brain

It is not possible to discuss the primates without saying at least something about their brains, for they are among the most highly developed in the animal kingdom; only those of the whales, porpoises and dolphins are anywhere near as advanced. As already seen, the movement forward of the eyes and the reduction in the size of the jaw musculature associated with an omnivorous diet has allowed expansion of the primate braincase, which means more space is available for the brain.

The part of the brain which has become most noticeably increased in size and complexity is the cortex. In the majority of mammals the surface of the cortex is smooth, but as the primate brain has enlarged so the cortex has become more and more folded and complex. It is perhaps obvious that the least complex cortical folding is to be found in the prosimians, for they are the more primitive primates. In the mouse lemurs, for example, folding is minimal, but it is possible to see a gradual increase in the more advanced lemurs and the simplest anthropoid brains have an even more folded cortex. The most complex brains are, of course, to be found in the great apes and humans, although one highly adaptable and successful group of monkeys, the New World capuchins, run them a close second. The capuchins' cerebellum and cerebral cortex are enlarged to the extent that, in relation to their body weight, these parts are larger than in any other monkey.

It is this enlargement of the brain which has allowed the primates to make maximum use of their environment and also to develop their characteristic advanced forms of social behaviour. Primates are animals with a high degree of locomotory skill, they are good at manipulating objects with their hands, they have very good eyesight, they often possess quite sophisticated oral com-

The lemurs, such as this white-fronted (*Eulemur fulvus albifrons*) male in Madagascar, have long snouts and wet noses associated with a well-developed sense of smell. The female is plain brown.

munication, and in humans they display intellect and emotion. The cerebral cortex has large areas associated with these features, with on the other hand a reduction in the area associated with the sense of smell.

The senses

A good sense of smell requires considerable volume within the nose in order to accommodate a large area of sensitive membranes which will trap the molecules of scent taken in as the animal sniffs. As a consequence of the gradual reduction in the length of the primate muzzle, space for these sensitive membranes has become less and less; accordingly, smell is the least developed of the primate senses. It is only in the prosimians that we find anything like the length of muzzle to be found in other mammalian groups.

Hearing is also a sense whose acuity has become gradually reduced in the majority of primates, with vision taking over the role in those which are diurnal. The primates with the most acute hearing are those which are nocturnal and which feed upon small animals such as insects, for they have to be able to detect the slightest movements of their prey in the stygian darkness of the forest nights. Prime examples are the bushbabies, the mouse lemurs and the aye-aye; the main indication of their aural acuity is their large external ears.

Sight plays a major role in the life of all primates including the nocturnal ones. The basic structure of the primate eye is the same as that for all other mammals, but whereas the majority of mammals have only black and white vision, colour vision is the norm in the diurnal primates. Colour vision is effected in the retina by the presence of special cells, distributed mainly towards the centre, called cones. In humans, apes and some monkeys there is a highly concentrated area of these cells, known as the fovea or yellow spot, right in the centre of the retina. In the region of peripheral vision on the retina the cones are replaced by different cells, the rods, which are sensitive at low light levels but see only in black, white and shades of grey. In nocturnal primates rods alone are present on the retina, and nocturnal prosimians have in addition a special layer behind the retina, the tapetum. This reflects back on to the retina light which has already escaped past it, thus increasing the total amount of light assimilated. If you search for these animals at night using a torch, light reflects back off the tapetum so brightly that their eyes resemble glowing coals.

The visual fields of all primates show at least some degree of overlap or convergence. This reduces the animal's overall field of vision, but on the other hand allows it to gain a better judgement of the position and distance of objects, such as food items, which are in front of it; in other words, it imparts some degree of stereoscopic vision. Because the eyes of prosimians are not as far forward on the face as those of apes and monkeys, prosimians have less well-developed stereoscopy than the latter. It is the extremely highly developed stereoscopic vision of the higher primates, coupled with their opposable thumb, that gives them their ability to manipulate small objects with great dexterity.

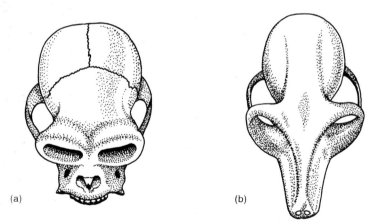

(a) (b)

Figure 4   A view from above the skull of an Old World monkey (a) and a lemur (b). The long muzzle of the lemur restricts the degree of forward vision, and thus stereoscopy, which is achieved by the monkey with its very short muzzle.

Manipulation of food and other objects requires a well-developed sense of touch. The hands and feet of the primates are well supplied with sense organs, especially on the tips of the digits. The 'fifth hand', the prehensile tail of the New World monkeys, is also well endowed with organs of touch.

An important role of the hands in primates is that of mutual grooming,

The South American douroucouli (*Aotus trivirgatus*), the only purely nocturnal simian, has the forward-facing eyes which give the monkeys and apes their stereoscopic vision. Photographed at Twycross Zoo.

Facial expressions play an important role in primate communication, though researchers are not as yet sure of the significance of the white eyelid display of the long-tailed macaque (*Macaca fascicularis*), demonstrated here by an old male in a Javan rainforest.

during which one individual picks meticulously through the fur of another in a search for virtually any loose items, such as bits of skin or pieces of food, which might be present. Grooming appears to be greatly enjoyed by both individuals, who swap jobs once the grooming of one of them is completed. The act of grooming seems to play an important part in primate social behaviour as a means of communication between members of a single social group.

Like other animals which live in social groups, primates communicate their moods and intentions to each other in a number of other distinctive ways. What these are will be dealt with only briefly here, for each system of communication will be considered in more detail in later chapters.

Figure 5 The so-called 'lip-flip' of the male gelada (*Theropithecus gelada*), in which the white teeth and gums of the upper jaw are revealed by eversion of the upper lip. At the same time, the scalp is retracted to reveal the white eyelids. No one is as yet sure of the meaning of this amazing facial expression.

Despite the fact that nearly all primates have a relatively poor sense of smell, scent signals are of some importance, especially in the prosimians. Both males and females of Coquerel's mouse lemur (*Mirza coquereli*), for example, produce at intervals a very strong scent which relays to other members of the species what sex they are and what reproductive state they are in. Sexual pheromones are produced by all other female primates to inform the males of the state of their reproductive cycle. Scent glands on various parts of the body of the New World monkeys communicate to other members of the group details of sex, age and social position, and are also used to delineate territorial boundaries between adjacent social groups. Some of the lemurs also possess special glands used in marking out their territories. As might be expected, the nocturnal night monkeys (*Aotus*) also rely heavily upon scent as a means of communication. Individuals of this species have a gland at the base of the tail accompanied by a special brush of hairs which is used to distribute the scent. Such scent glands are scarce in Old World monkeys and the apes, for whom visual communication seems to play a greater role than does communication by olfactory means.

The most obvious forms of visual communication in primates are the colours and patterns found on the face, body and tails of many species. These are used to convey to others information about which species a particular individual belongs to or, in the case of sexually dichromatic species, what sex the individual is. Body posture, fluffing up of fur, the position in which the tail is held and facial expressions are also all used in visual communication. Facial expressions, of course, play an important role in human communication, but these are not unique to us. Thus male geladas can present a quite remarkable facial expression by flipping back their upper lip to show the naked muzzle beneath, as well as retracting the scalp to reveal their white eyelids.

Mention was made earlier how during female oestrus in Old World monkeys areas of skin swell up and become brightly coloured, a clear indication to the males that the females are in mating condition. To a certain extent this form of communication replaces the olfactory signals used by the other primate groups. In the majority of those monkeys which employ this kind of visual communication it is skin in the genital region which becomes swollen and brightly coloured. The exception is the gelada, which spends more of its life sitting on its rear end than do other primates; the coloured sexual skin is therefore sited on its chest instead of on its rear.

Visual communication is clearly important when a group of primates are in close proximity to one another; but between different social groups of nocturnal species, and when individuals of these species are foraging in the dense foresty canopy, auditory communication plays a very important role. The amount of sound produced depends upon the situation. As the members of a group of golden bamboo lemurs, for example, feed in dense bamboo forest they maintain contact by emitting regular, muted sigh-miaows. On the other hand at dusk, and occasionally even during the morning, the male launches forth, with his head thrown right back, into his 'great call', whose volume of sound is truly amazing for an animal only the size of a small cat. These calls in primates are often associated with the demarcation of territorial boundaries and are intended to warn off would-be intruding groups.

One call that is common to most tree-dwelling primates is the signal given when an eagle or other raptor threatens. The use of alarm calls to alert other individuals to the presence of a predator has been considerably extended in the mainly terrestrial vervet monkeys from the African savannahs. They have evolved a series of calls which are literally words, in that they inform others of which particular predator is present. We know this to be so, for a specific call will always elicit a fixed response from the monkey who hears it (see Chapter 8).

CLASSIFICATION OF PRIMATES

In order to understand this book fully it will be useful to take a closer look here at those animals in existence today which actually make up the order Primates. The suborder Prosimii contains the more primitive primates – the lemurs, lorises, pottos and bushbabies. Externally they have relatively long muzzles terminating in a naked, moist snout, the rhinarium. Their whiskers are well developed and so, in most of them, are the external ears. The larger species of lemurs are diurnal, whereas the smaller species of lemurs and the other prosimians are generally nocturnal and have large eyes in relation to the size of their faces. They differ in tooth numbers from the higher primates and the lower jaw, the mandible, is made up from two separate bones joined by cartilage.

The suborder Anthropoidea contains the monkeys, gibbons, apes and man. This group has short muzzles, with the exception of the baboons; the nose is dry and whiskers are poorly developed. The external ears are very small, since

The slender loris (*Loris tardigradus*) is, as here, most likely to be seen in a zoo since it is a purely nocturnal creature and difficult to find in the wild.

the Anthropoidea are almost without exception diurnal and rely heavily upon sight. The mandible consists of a single bone.

Apparently intermediate between these two suborders is the suborder Tarsioidea. Like the anthropoids, the tarsiers have a short muzzle and a dry nose, but the mandible consists of two separate bones.

The suborder Prosimii is subdivided into two infra-orders, the Lorisiformes and the Lemuriformes. The Lorisiformes contain a single family, the Lorisidae, which incorporates the lorises from Asia, the potto and the angwantibo from Africa and the bushbabies or galagos, also African. The classification of the Lemuriformes is slightly more complicated in that they comprise two superfamilies, the Lemuroidea and the Daubentonioidea. The latter contains a single species, namely the weird aye-aye (*Daubentonia madagascariensis*) from

 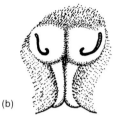

Figure 6 The dry nose and complete upper lip of a simian (a) and the naked wet rhinarium and split lip of a prosimian (b), which serve as one of the means of separating these two primate groups.

Madagascar. Within the Lemuroidea there are two families differentiated by their tooth counts; the Lemuridae, containing the true lemurs and the dwarf lemurs, with a total of thirty-six teeth, and the Indriidae, containing the indri and the sifakas, with only thirty teeth.

The suborder Anthropoidea is subdivided into two infra-orders, the Platyrrhini and the Catarrhini. The platyrrhines are the New World monkeys, recognizable externally by their broad noses and sideways opening nostrils. The catarrhines are the Old World monkeys, the apes and humans, with their narrower noses and downward-facing nostrils. They also differ from the platyrrhines in that some of them have ischial callosities on their rear ends, on which they sit. Within the New World monkeys there are two families, the Callitrichidae, the marmosets, with thirty-two teeth and the Cebidae, the remainder, with thirty-six teeth.

The Catharrhini are further separated into four families. The Cercopithecidae are the Old World monkeys from Asia and Africa, the Hylobatidae are the gibbons and siamangs from Asia, the Pongidae include the gorilla and chimpanzees from Africa and the orang-utan from Asia, and the Hominidae include man.

EVOLUTION OF THE PRIMATES

When tracing the evolution of the order Primates a seemingly insurmountable problem arises: the lack of available fossil material. The reason relates to the forest-dwelling way of life that has been followed by most primates, for fossilization rarely occurs under these conditions. Of the fossils of primates so far discovered more than 65 per cent are fragmentary remains – usually just a few bits of bone or teeth. As a result less than 35 per cent of the material so far obtained can be assigned with any certainty to the order Primates.

The first recognizable mammals appeared as long as 210 Ma (Ma = millions of years ago), and the first placental mammals were present about 70 million years later. All of the substantial primate fossils come from the Cenozoic period, that is from the period between 66 Ma and the present day.

Before that time the primates seem to have split into two main streams: one of them led to a number of groups, all of which became extinct many millions of years ago and do not really concern us here; the second led to the modern primates.

From near the beginning of this period comes a fossil which is generally accepted as lemuroid and which may be close to the ancestor of the living lemurs. *Notharctus* lived in what is now North America and, although it was very lemur-like in appearance, its dentition was primitive – it retained four premolars in each half jaw, and there was little or no indication of the modification of the canines and incisors to form the dental comb of the present-day lemurs. We know nothing of what happened to the lemurs between the Eocene and the Pleistocene, when their fossils are to be found once more. Fossils of tarsioids have also been found from the Eocene period, but the lorisoids do not appear until the Miocene of Africa some 20 Ma. In these fossils the dental comb is well developed and presumably evolved at some time between the Eocene and the Miocene. Quite what the relationships between the Eocene lemuroids and tarsioids are is not clear, nor is the way they relate to the Miocene lorisoids, though the fossil family Onomyidae from North America may be close to the ancestral stock for all three of today's main primate groups.

Although the prosimians are known from the Eocene, there is only one anthropoid fossil from this time, and even that is doubtful; it is not until the early Oligocene, 28–35 Ma, that the first definite remains of this type make their appearance. These fossils come from Egypt and are thought to represent the line from which the present-day Old World monkeys, the apes and humans, have come. There are two hypotheses as to the origins of the New World monkeys. One is that they originated in the northern hemisphere, then migrated through North America and eventually reached South America. The second is that they migrated from Africa into South America. Which of these seems the more likely?

Since the continents are not fixed in position but are drifting around on the surface of the earth, the relative positions of Africa, North America and South America were very different 30 Ma from what they are today. A very large gap existed between the two Americas, and Africa was in fact much closer to South America than the latter was to North America. If we consider that the monkeys might have reached South America by island-hopping or on floating islands of vegetation coming out of the large rivers on the west coast of Africa, then obviously the second hypothesis is the more likely. Indirect evidence for a common ancestor for all the simians comes from studies on the structure of their proteins and the arrangement of their chromosomes and gene patterns. What this does not tell us, however, is whether this common ancestor was indeed one of the early African simians or a prosimian.

Of the anthropoids the South American platyrrhines still retain a number of primitive features: for example, with the exception of the uakaris (*Cacajao*) there is never any reduction of the long tail and there is little opposability of the thumbs. They all have three premolar teeth on each side of each jaw, whereas even in the oldest-known Old World monkey fossils these have been reduced to

The pygmy marmoset (*Cebuella pygmaea*), photographed here in Twycross Zoo, shows fairly primitive primate characteristics. Marmosets are one of the few South American monkeys of which any fossil evidence has been discovered.

two. Fossil evidence of New World monkeys is somewhat sparse. Teeth from a marmoset have been found dating from the middle Miocene of La Venta in Colombia, and a cebid has also been found dating from the Miocene period. *Homunculus*, as it has been named, belongs within the same group as the douroucouli (*Aotus*), which retains the thick, woolly coat and small brain of primitive species. *Aotus* is similar to the marmosets and may in fact be quite close to the basal stock of the platyrrhines.

The oldest-known fossil catarrhines all originate from the Fayum Depression in Egypt. A number of species have been described, all from the Oligocene, some of them probably close to the basal stock of the modern catarrhines. *Parapithecus* can be considered as primitive since it still had three premolars, whereas *Oligopithecus* from 30–35 Ma had the two premolars of today's Old World species. *Propliopithecus* from 28–30 Ma and *Aegyptopithecus* from 28 Ma both had ape-like dentitions. At what point these gave rise to the modern catarrhine families – if they even did so – we are not sure, for the earliest fossils that can be assigned to the modern catarrhine groups are from the Miocene, some 10 million years later.

The earliest cercopithecoids date from 18 Ma, from fossil deposits in North Africa. From about 3 million years later are two fossils from Kenya which are close to the basal stock of the two cercopithecid subfamilies, for one is of a leaf-eating colobine type and the other resembles an omnivorous cercopithecine. By 10 Ma, however, the colobines as a group were well developed and one of these, *Mesopithecus*, a langur-like monkey from the Pliocene of Greece, is known from a virtually complete skeleton. The earliest appearance of the colobines in Asia is in fossils dated as from 7 Ma in the late Miocene.

The cercopithecine monkeys also lived in Europe at one time, having arrived there from their original home in Africa, and fossil macaques have even been found as far west as the British Isles. The arrival of the cercopithecines in Asia may have been somewhat later than that of the colobines, for the earliest-known fossil cercopithicine from that area is as relatively recent as 3 Ma. Within Africa the latter evolved during the Pliocene into the three main groups – the geladas, the baboons and their kin, and the guenons – and fossil remains of all three have been found.

The higher apes may well have evolved from *Propliopithecus* or a close relation. Very close to this animal was *Pliopithecus*, which was very similar in most of its characteristics to the living gibbons. From the middle Miocene of both Africa and Asia come fossils of two genera of pro-apes, *Ramapithecus* and *Sivapithecus*. It is believed that the orang-utan, which in the Pleistocene was also found in mainland Asia, may have evolved from the Asian line of one of these. The probable common ancestor of the African apes and of humans has long been considered to be *Proconsul* from the early Miocene of East Africa. There is no fossil evidence of any early form of chimpanzee or gorilla, and we do not yet know at what point humans diverge from the great ape line. There is an increasing fossil record of men-like apes and ape-like men, but the overall story of the evolution of humans does not lie within the scope of this book.

Having considered in outline in this chapter the classification of the extant Primates and their possible origins, we will now look at each of the families in more detail.

Chapter 2
The Prosimians and New World Monkeys

The Prosimians

THE LORISIDAE
Within this family there are two subfamilies; the first is the Lorisinae, which contains the lorises proper. The slender loris (*Loris tardigradus*) inhabits forested areas of southern India and Sri Lanka, where it feeds mainly on insects. Like all members of this family, it is nocturnal in habits and it hunts its prey by proceeding quadrupedally in a slow and deliberate manner through the trees in which it lives. The slow loris (*Nycticebus coucang*) is widespread in south-east Asia, including Assam on the mainland of the Indian sub-continent, where it is nocturnally active in the tropical rainforests. It too moves around in the slow, deliberate manner characteristic of the family, but as well as insects it feeds on leaves, fruit, birds' eggs and even young birds from the nest when it comes across them.

Africa boasts just two species from this family, both of which live in trees, are nocturnal and move around in the manner typical of the lorises. The angwantibo (*Arctocebus calabarensis*) is restricted to forested areas between the Zaire and Niger rivers in central Africa. Here it lives in the shrub layer where it feeds mainly on insects with a supplement of fruits when available. The range of the angwantibo overlaps that of the potto (*Perodicticus potto*), which is also found in Kenya and extends into West Africa as far as Sierra Leone. Its feeding range, however, does not overlap with that of the angwantibo, for the potto feeds mainly on plant material, especially fruits, with a supplement of animal protein in the form of insects which it finds in the forest canopy. The potto has a short tail and is somewhat larger than the tailless angwantibo.

The second subfamily within the Lorisidae, the Galaginae, contains the bushbabies or galagos. These are all of African origin and are much more agile creatures than their slow-moving loris cousins. The actual number of species recognized varies from six to eleven, depending upon the authority concerned; the most recent symposium (1986) upon the subject recognizes eleven species in three separate genera.

The animals which most people recognize as the typical bushbaby belong to the genus *Galago*. The lesser or Senegal bushbaby (*Galago senegalensis*) is found in much of the forest and savannah woodland lying south of the Sahara and north of the Zaire basin in central Africa, where it feeds upon plant gum and invertebrates. It also occurs throughout Kenya, Tanzania and Uganda. It is absent from the central African rainforest areas, where it is replaced by the following species. South of the Zaire basin and into southern Africa down to

The potto (*Perodicticus potto*), seen here in Cincinnati Zoo, is typical of the slow-moving, nocturnal lorisiforms from the tropical forests of Africa and Asia.

the Orange River lives the mohol (*G. moholi*), only recently split off from *G. senegalensis*; this species also feeds upon gums and invertebrate animals. Also once included as a subspecies of the latter is *G. gallarum*, the Somali galago, which is found in northern Kenya, southern Somalia and south-eastern Ethiopia. *G. gallarum* is reported as feeding upon insects and the fruit of various trees subject to seasonal availability. They are typical leapers, with long hind-limbs, and, despite their small size (160 mm [6.5 in] for the head and body of the lesser bushbaby), they can perform vertical jumps in excess of 2 m (7 ft). That there were three species in existence, rather than just *G. senegalensis*, became evident when it was found that their ranges overlapped but that they maintained their identities without interbreeding. All three species are nocturnal and have enormous eyes and large, membranous ears which can be folded up against the head for protection as they forage through the trees.

The genus *Galagoides* contains four species. Allen's bushbaby (*Galagoides alleni*) is slightly larger than the lesser bushbaby, with a head and body length between 200 and 280 mm (8–11 in) as against 140–210 mm (6–8 in) for the latter. Allen's bushbaby also has a somewhat sharper and longer muzzle. It shares its range with two other species, the western needle-clawed and Demidoff's bushbabies. Allen's bushbaby is mainly vegetarian, feeding especially upon gum, but it does take a few insects too. The dwarf or Demidoff's bushbaby (*G. demidoff*) inhabits primary and secondary forest in Nigeria, central Africa and parts of Kenya, Tanzania and Malawi. As its common name implies, it is the smallest species of bushbaby, but it resembles the others in apearance. Its feeding habits are similar to those of the lesser bushbaby – mainly insects and gum, with some fruit when available.

Once considered to be a subspecies of the dwarf bushbaby is Thomas's bushbaby (*G. thomasi*). It is, however, larger than *G. demidoff*, and where their ranges overlap they do not interbreed. *G. thomasi* has a very disjunct range since it is basically a montane species, found mainly on mountains around the Zaire basin. Little is known of its feeding habits. The fourth *Galagoides* is *G. zanzibaricus*, which is found in East Africa from the coastal regions of Somalia, Kenya and Tanzania into Mozambique and inland in Malawi and south-east Zimbabwe, as well as on the island of Zanzibar. Intermediate in size between Allen's bushbaby and *G. thomasi*, it feeds mainly on invertebrates and fruit.

The greater bushbaby (*Otolemur crassicaudatus*) is, as its name implies, the largest of the group, about the size of a rabbit. It is found in central Africa, where it inhabits woodland savannah and open forest between the Equator and the Tropic of Capricorn. It resembles the other galagos in appearance except for its very densely bushy tail, from which its species name is derived. *O. crassicaudatus* feeds mainly upon fruit and gums, with a supplement of insects when it can catch them. Although somewhat less agile than the other galagos, on the ground it can move at considerable speed in a kangaroo-like manner. Garnett's or the small-eared greater bushbaby (*O. garnettii*) is an inhabitant of the coastal regions of East Africa from Somalia down to Tanzania, and it will almost certainly be discovered in similar habitats along the coast of Mozambique. It also occurs on the island of Zanzibar. Its range overlaps that of the

Zanzibar bushbaby but it is considerably bigger: as adults, their sizes do not overlap. *O. garnettii* feeds upon invertebrates and fruit.

Closely related to these two species are the two species of needle-clawed or needle-nailed bushbabies sometimes placed in the genus *Euoticus*. Their common name comes from the unique construction of their nails, which have a central keel narrowing down to a very pointed tip. The latter allows these animals to grip the bark of large forest trees and climb both vertically up and head first down the trunk. In the canopy they feed upon plant gums, insects and some fruit. Both species are typical bushbabies in appearance. The western needle-clawed or elegant bushbaby (*Galago elegantulus*) occupies a range between the Niger and Zaire rivers in central Africa. The eastern needle-clawed bushbaby (*G. matschiei*) is restricted to a much smaller area along the eastern border of Zaire from Lake Mobuto to Lake Tanganyika.

THE LEMURIDAE

The lemurs are restricted to the large island of Madagascar and to the Comoro Islands in the Indian Ocean, though the present feeling is that lemurs from the latter were originally introduced there by humans. They vary in size from the tiny mouse lemurs, which as their name implies are the size of a mouse, to the largest species such as the ring-tailed lemur, which is the size of a domestic cat. The smaller species are nocturnal in behaviour, the larger species diurnal.

The mouse lemurs such as the rufous (*Microcebus rufus*), seen here at night in a Madagascan rainforest, are the smallest of the primates.

The Lemurinae

The subfamily Lemurinae, the so-called true lemurs, contains four genera, *Lemur*, *Hapalemur*, *Lepilemur* and *Varecia*. Up until very recently the genus *Lemur* was considered to contain six species, of which the best known is almost certainly that great show-off, the ring-tailed lemur (*Lemur catta*). Some up-to-date research has, however, cast a different light upon the relationship between *L. catta* and the other five species in the genus. It would appear from chromosome studies and the presence of glands on the arms that *L. catta* is much more closely related to the members of the genus *Hapalemur* than it is to the other *Lemur* species. Accordingly it has been decided that the *Lemur* species other than *L. catta* should be placed in the genus *Eulemur*, with the mongoose lemur (*Eulemur mongoz*) designated as the type species. Bearing these changes in mind, we can now consider the characteristics of the various true lemurs.

The ring-tailed lemur is immediately recognizable on account of its long black and white ringed tail and its black eyes, nose and mouth contrasting with its white face. As already mentioned, it may be distinguished from members of the genus *Eulemur* by the presence of special scent glands situated on the forearm and the inner side of the upper arm. Ring-tails move around the forests of south-western Madagascar in groups of up to twenty-four individuals, feeding upon plant material such as fruits and seeds. Although they generally move around the trees in which they live in a quadrupedal fashion, they do have relatively long hind-limbs which enable them to indulge in a certain amount of leaping from branch to branch.

The genus *Eulemur* contains five species, of which the black lemur and the brown lemur have clearly distinguished subspecies. The black lemur (*E. macaco macaco*) is endemic to the humid forests of north-western Madagascar and it is also found on the islands of Nosy Bé and Nosy Komba. This species shows a well-marked sexual dichromatism, with the males completely black while the females are reddish brown with white tufts on the ears. The second subspecies, Sclater's lemur (*E. macaco flavifrons*), was only rediscovered in the wild in 1983 in a strip of coastal forest in north-west Madagascar. The males resemble the black lemur, though their black colouring is less intense and has a brownish reflection particularly on the underside. The male also has a crest formed from a series of short, erect hairs on top of the head. The females are more orange than females of the black lemur, with a gradation from darkest along the middle of the back to almost white on the underside; they also have a pale band on the forehead and round the eyes on to the cheeks, but ear tufts are absent. Whereas the eyes of the black lemur are yellow or reddish, those of Sclater's lemur are blue-green.

Apart from the brown lemur proper (*Eulemur fulvus fulvus*) there are six other recognized subspecies of *E. fulvus*. One of these, *E. f. mayottensis*, occurs on the island of Mayotte in the Comoro Archipelago. Of the others, the red-fronted lemur (*E. f. rufus*) is found in areas in both the east and west of Madagascar; the white-fronted lemur (*E. f. albifrons*) comes from the humid eastern forest regions; the collared lemur (*E. f. collaris*) occurs in south-western Madagascar; Sanford's lemur (*E. f. sanfordi*) is found only around Montagne d'Ambre in the

A group of brown lemurs (*Eulemur fulvus fulvus*) participating in a daytime siesta in tropical dry forest in Madagascar. This is the only *fulvus* subspecies in which males and females are similar.

north; and the white-collared lemur (*E. f. albicollaris*) lives in an area of the eastern humid forest. *E. f. fulvus* has a disjunct distribution, having individual centres of population in the north-west, north and east of Madagascar.

The various subspecies show some variation in their social groupings and diet, dependent to some extent upon where they are found. Some feed, for example, upon leaves alone, while others supplement this diet with fruit and flowers. *E. f. mayottensis* and possibly some of the other subspecies exhibit diel activity – that is, they are as active at night as in the day.

The mongoose lemur (*E. mongoz*) is the second species of lemur found outside Madagascar, since it also lives on two of the Comoro Islands. On the mainland it is found in the north-western forests, where it lives in family groups and feeds on a very specialized diet of the flowers of just five species of plant. The relatively little-known red-bellied lemur (*E. rubriventer*) comes from the fast-disappearing eastern forests of Madagascar. They live in small family or all-male groups, feeding mainly on fruit and, when this is not available, flowers and leaves. Finally within the genus *Eulemur* is the crowned lemur (*E. coronatus*), which is found in both dry and moist forests in the north of Madagascar, where it lives in relatively large groups containing both sexes. As with *E. macaco* and some varieties of *E. fulvus*, the sexes are noticeably different in appearance. The males are a rich reddish brown with a fine haze of black; the eye-ring and muzzle are pale and whitish, the cheeks and 'eyebrows' rufous, and on top of

The silvery grey female crowned lemurs (*Eulemur coronatus*) are considered by many people to be more attractive than the brown males. This individual was photographed in Madagascan tropical dry forest.

the head is a solid black 'crown' which projects forwards in a triangle between the ears. The females are a uniform steel-grey except for a V-shaped rufous crown on the forehead. The underside in both sexes is off-white.

Up until very recently the genus *Hapalemur* contained only two species; now there are three. All of them live in the bamboo forests of Madagascar, and these plants form their principal food item. Because of the nature of the habitat in which they live, they are all adapted for vertical clinging and leaping from one bamboo stem to another. The grey gentle or bamboo lemur (*H. griseus*) is the most widespread of the three, living in both the east and north-west of Madagascar. Four subspecies are recognized, each from its own individual area of the island. The eastern grey gentle or bamboo lemur (*H. g. griseus*) is found in the humid forests of the east; the larger grey gentle or bamboo lemur (*H. g. alaotrensis*) lives in the reed beds of Lake Alaotra and also the neighbouring marshes; the western grey gentle or bamboo lemur (*H. g. occidentalis*) is known only from two isolated areas in the west; and a new subspecies, *H. g. meriodinalis*, from the south-east coast was described in 1986. These diurnally active lemurs usually live in small groups and, although bamboo is their main food, they have been seen to take other plant food.

The broad-nosed or greater gentle or bamboo lemur (*H. simus*) is now found in only a small area in the south-east of Madagascar. It was once more widespread, but its fate is linked with the disappearance by clearing of its main source of food, the giant bamboo – 90 per cent of *H. simus*'s diet consists of the plant's pith. The species still survives in the bamboo forests at Ranomafana, where it is found in company with *H. griseus* and the most recently described lemur, the golden bamboo lemur (*H. aureus*). The latter was first recognized as something different in 1985, and was described as new to science in 1987. In size it comes between the other two species, but it is a more attractive animal than either. The blackish brown muzzle is bordered by biscuit-coloured cheeks and eyebrows, while the top of the head and the span of the back extending down on both sides are a light chocolate-brown rather than gold, which, despite the animal's name, is strictly speaking confined to the chest fur. It lives in groups of two to four among the stems of giant bamboo, feeding on the leaves and young stems, although it will also take bamboo grass and certain fungi when they are available.

Once included in *Lemur* but now in a genus on its own because of its unusual female reproductive physiology is the ruffed lemur (*Varecia variegata*), the largest of the true lemurs. There are two recognized subspecies, the black and white ruffed lemur (*V. v. variegata*), which comes from most of the eastern strip of humid forest, and the red ruffed lemur (*V. v. rubra*), which is found only on the Masoala peninsula. Very little is known about the ruffed lemur in the wild; it appears that they usually move around quadrupedally in pairs, and they have been observed feeding on fruit.

Finally within the true lemurs there is the sportive or weasel lemur (*Lepilemur mustelinus*). At times the genus *Lepilemur* has been subdivided to form at least seven species of sportive lemur, but there does not seem at present to be any firm basis for doing so. Suffice it to say that the species shows some degree of variation throughout its range. Not a great deal is known about this nocturnal lemur, which inhabits both humid and dry forests throughout much of Madagascar where it apparently feeds mainly on leaves. It moves through its forest habitat by vertical clinging and leaping.

The Cheirogaleinae

This subfamily, known collectively as the dwarf lemurs, contains four genera all the members of which are nocturnal in habit. The mouse lemurs of the genus *Microcebus* include two species. The grey or lesser mouse lemur (*M. murinus*) averages a mere 125 mm (5 in) head and body length combined, making it one of the smallest primates. It is abundant throughout the forests of western, southern and south-eastern Madagascar, where it feeds omnivorously upon fruit, flowers and some leaves, and on insects when it comes across them. This species is replaced in eastern Madagascar by the equally abundant brown or rufous mouse lemur (*M. rufus*), which has similar habits to *M. murinus*. There is a third species, Coquerel's mouse lemur (*Mirza coquereli*), which is nearly twice the size of *M. murinus* and comes from the fast disappearing forest areas of western Madagascar. Like the other two species it is omnivorous, feeding on a mixture of fruit, flowers and insects and at certain times of the year insect exudates. All three species have relatively short limbs and move around the trees and shrubs on which they live in a purely quadrupedal fashion. Females and young tend to occur in groups, often sharing a daytime sleeping nest, whereas males tend to sleep alone or at the most in pairs.

The two species of dwarf lemurs of the genus *Cheirogaleus* live in very different environments. The greater dwarf lemur (*C. major*) is an inhabitant of the eastern rainforest areas, where it feeds on ripe fruits, nectar and pollen. The fat-tailed dwarf lemur (*C. medius*) lives in the dry forests of the north, west and south-west of the island, where its diet depends upon the availability of various foods. It is mainly frugivorous and will also take nectar; but, unlike the previous species, which has never been seen to take them, it also feeds upon insects. The *Cheirogaleus* lemurs move quadrupedally, usually singly, though they are known to occur at fairly high population densities – as many as 350 individuals per 100 ha (roughly 85 per 100 acres) for *C. medius*.

The remaining cheirogaleine genera each contain just a single species. The fork-marked lemur (*Phaner furcifer*) has a discontinuous distribution, mainly in western Madagascar but with small isolated populations in the north, east and south. Its alimentary canal, teeth, tongue and nails are modified for its specialized diet of plant gums, though it has also been seen to take flowers, fruit and insects. It is a solitary animal and moves quadrupedally around the trees and shrubs on which it feeds. The final species, the hairy-eared dwarf lemur (*Allocebus trichotis*), was until recently thought to be extinct, since no living individual had been seen since 1965. At one time it had a fairly wide distribution in the eastern humid forests of Madagascar, and there appears to be no logical reason for its decline. Very recently this little-known lemur was rediscovered in the area of its original locality in rainforest near Mananara. Very little is known about it.

THE INDRIIDAE

Within this group are contained three genera which are separated off from the Lemuridae on account of their reduced number of teeth. The indriids have one premolar tooth less on each side of both jaws, and they have also lost a pair of incisors from the lower jaw. Despite this difference they are typical lemurs in external appearance. Structurally they have the long hind-limbs and shorter front limbs of vertical clingers and leapers. The indri (*Indri indri*), the largest of the lemurs, is found in certain parts of Madagascar's north-eastern forest areas where it lives in small family groups. It resembles the sifakas in general appearance, but unlike them has only a very short tail. Since it is a heavy animal it tends to be found moving around on the more robust branches of the trees from which it obtains its food of fruit, shoots and young leaves.

Verreaux's sifaka (*Propithecus verreauxi*) occurs in forests down the western

The beautiful Verreaux's sifaka (*Propithecus verreauxi verreauxi*) is a member of the family Indriidae. This young individual is curious, peering at the photographer from close range in Madagascan dry forest.

side and in the south of Madagascar. Four subspecies have been described, each with its own distinctive colouring. Verreaux's sifaka (*P. v. verreauxi*) is found in the south and south-west of the island, while Coquerel's sifaka (*P. v. coquereli*) comes from the forests of the north-west. The completely white Decken's sifaka (*P. v. deckeni*) lives in the central area of western Madagascar, and the crowned sifaka (*P. v. coronatus*) occupies an area between Coquerel's and Decken's sifakas. It is, however, possible if not probable that the latter two are just separate phases of a single subspecies; further field studies may provide the answer.

The diademed sifaka (*Propithecus diadema*) is distributed down the eastern side of Madagascar and has five described subspecies. Perrier's diademed

The eastern avahi (*Avahi laniger laniger*) is the smallest of the indriids and is purely nocturnal. It spends the day, as here, grasping a tree trunk or perched in the crook of a branch. These two have been photographed during their daytime rest in dense cover in a Madagascan rainforest.

sifaka (*P. d. perrieri*) is completely black and comes from the extreme north-east of the island. The silky diademed sifaka (*P. d. candidus*) also comes from the north-east, but from further south than *P. d. perrieri*. Moving south down the east side of Madagascar we then encounter in turn the other three subspecies, the typical diademed sifaka (*P. d. diadema*), the Milne-Edwards diademed sifaka (*P. d. edwardsi*) and finally the black diademed sifaka (*P. d. holomelas*), which, like Perrier's sifaka, is wholly black. Recent work on the last two subspecies has led to the conclusion that they should both be included under *P. d. edwardsi* as a single subspecies. Like the indri, the sifakas move around the forest in family groups where they feed upon a diet consisting mostly of leaves and fruit, though they will also take flowers and bark.

In 1974 a population of *P. diadema* was reported from an area of forest in which it had not previously been recorded. This animal was provisionally assigned to the nearest recorded subspecies, *P. d. candidus*, despite the fact that they differed substantially in several ways. It was not until 1987, when their habitat came under serious threat, that a number of them were taken into captivity for breeding purposes. It was then realized that they were considerably smaller than any of the other *P. diadema* subspecies and, unlike *candidus*, they had short white fur rather than long silky white fur. They also had furry ears with long hair forming tufts which extended well beyond the ear tips. Further differences, including a chromosome count and a voice different from that of the other two *Propithecus* species, led to the realization that this was in fact a new species, Tattersall's sifaka (*Propithecus tattersalli*), named after the man who discovered it, Ian Tattersall. It is restricted to a small area of primary dry forest in the far north of Madagascar, where its distribution is very fragmented. Like the other sifakas it is a vegetarian; otherwise little is known of its habits in the wild.

The final member of the Indriidae is the smallest, the woolly lemur (*Avahi laniger*). This species differs in another way from the rest of the indriids in that its members are strictly nocturnal. They live in small family groups, and their diet is much the same as that of the sifakas. Two subspecies are recognized: the eastern woolly lemur (*A. l. laniger*) is found throughout most of the eastern area of humid forest, while the western woolly lemur (*A. l. occidentalis*) occurs only in a very small area in the north-west of the island.

THE DAUBENTONIIDAE

This family now contains just a single species, the strange aye-aye (*Daubentonia madagascariensis*), though a second species, *D. robusta*, appears to have existed on the island up to less than a thousand years ago. The aye-aye has always been thought of as rare and endangered, and was at one time believed to be on the verge of extinction. It now appears that it may be more common than originally thought, and that because of its retiring nocturnal habits it is under-recorded. The aye-aye cannot be mistaken for any other lemur – some people liken its appearance to that of a cat. It is unusual in that it has claws rather than nails on all of its digits except the big toes. It feeds on fruit and the wood-boring grubs of insects, which it extracts with its elongated middle fingers. These are also used to scoop the flesh out of coconuts after it has gnawed a hole into them with its specially adapted incisors. It moves around quadrupedally in the trees, and with the aid of its sharp nails can climb up their vertical trunks with ease. During the day it sleeps in large nests fairly high above the ground; little is known about its social habits.

The aye-aye (*Daubentonia madagascariensis*) is the only living member of its family. On this individual, seen here at night in rainforest in Madagascar, note the large ears, with which it picks up the sounds of insect larvae in burrows beneath the bark of trees, and the long, thin middle finger with which it extracts these grubs.

Figure 7 A tarsier, showing its large eyes, associated with a nocturnal lifestyle, and its large ears, with which it detects the sounds made by the small animals such as insects on which it preys.

THE TARSIIDAE

The three extant species of tarsier are small, nocturnal animals confined to various islands in what used to be referred to as the East Indies. They are all very similar in appearance with their large eyes, large membranous ears, noticeably long hind-limbs and very long, almost naked tail. An odd point about the tarsiers is that they are unable to move their eyes within the sockets; they compensate for this by being able to rotate their head through 180 degrees. Tarsiers appear to move around in small family groups, using vertical clinging and leaping with their very long hind legs to move from tree to tree in search of their exclusively carnivorous diet of insects, other invertebrates and small vertebrates such as lizards if they find them. The three species are Horsfield's tarsier (*Tarsius bancanus*) from Borneo, Sumatra and other adjacent islands; *T. spectrum*, the spectral tarsier from Sulawesi and surrounding islands; and finally the Philippines tarsier (*T. syrichta*).

The New World Monkeys

THE CALLITRICHIDAE

This family contains two subfamilies which are considered to be the most primitive of the New World monkeys. They are all diurnal forest-dwelling monkeys.

45

The Callitrichinae

This subfamily includes the marmosets proper and the tamarins, which between them account for fifteen distinct species. They have an interesting distribution, for the marmosets are found exclusively south and east of a line formed by the Amazon and one of its main tributaries, the Madeira, while the tamarins are found exclusively to the north and west of this line. They do, however, overlap in one small area just south of the mouth of the Amazon, where the silvery marmoset and the midas or red-handed tamarin live side by side.

The true marmosets in the genus *Callithrix* are small monkeys with a head and body length of roughly 210 mm (8.5 in) and the tail about half as long again. The silvery marmoset (*C. argentata*) has three clearly recognizable races, which range in marking from being dark all over, through white with a dark tail, to white with a cream tail. The dark forms are easily distinguished from the other species by the naked ears. The Santarem marmoset (*C. humeralifer*) has tufts of hair arising from the ear rims, while the remaining species have tufts and plumes of hair arising from around the ear region. The number of other species of *Callithrix* depends upon which authority is consulted: some primatologists consider there is just one other variable species, while others believe that this should be separated into a number of distinct species. On the basis that there are a number of distinct individual populations which do not come into contact in the wild, the second option is being adopted here, giving a further five species.

All five species are contained in the *jacchus* group and include the common marmoset (*C. jacchus*), which is marbled black and grey in colour with a tail of black and grey rings. Most of our knowledge of marmosets comes from this species, but less is known of the daily lives in the wild of the other members of the group. The four other species recognized as such are the buffy-tufted-ear marmoset (*C. aurita*), the buffy-headed marmoset (*C. flaviceps*), the white-fronted marmoset (*C. geoffroyi*) and the black-pencilled marmoset (*C. pencillata*).

Marmosets inhabit most types of forest, where they feed upon insects, fruit and tree exudates. They cause the latter to be produced by chiselling holes into the bark of various trees, using their lower incisors and canines. Marmosets move around quadrupedally, and the presence of claws rather than nails on their digits enables them to climb up or dodge behind vertical trunks. They live in family groups of a mother and father and their offspring of different ages. They breed very easily in captivity, and as a result are commonly used in biological and medical research laboratories.

The callitrichines include the world's smallest monkey, the pygmy marmoset (*Cebuella pygmaea*), which fits neatly on to an average human hand, the head and body being 130 mm (5 in) long with the tail just over half as long again. *C. pygmaea* inhabits the upper reaches of the Amazon basin in Brazil, Colombia, Peru, Ecuador and northern Bolivia. It lives in company with tamarins over its range, where it is an inhabitant of rainforest. Here its brown fur, with each hair made up of alternate black and buff bands, camouflages it well as it darts along from branch to branch. Like the *Callothrix* marmosets it uses its lower incisors and canines to cut into bark to obtain its major food supply of plant exudates, which it supplements with fruit, leaves and insects. Pygmy marmosets live in family groups and, like the other marmosets, move around quadrupedally, although they have been observed to indulge in some vertical clinging and leaping on occasions.

The largest number of callitrichine species occurs in the tamarins of the genus *Saguinus*, which as already noted live north and west of the River Amazon. One species, however, reaches as far north as Panama and Costa Rica in Central America. Externally they are not dissimilar in appearance to the marmosets, but on average they are slightly larger. The main difference between the two lies in the structure of the lower jaw, which in the marmosets is V-shaped, with the canines and incisors equal in length, whereas in the tamarins it is U-shaped with the canines longer than the incisors.

The claws instead of nails, typical of the marmosets, show up clearly in this black-pencilled marmoset (*Callithrix pencillata*) in gallery forest in Brazil.

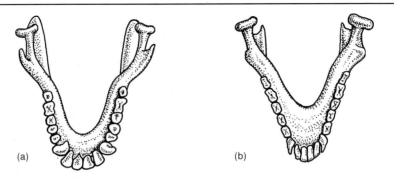

Figure 8 The lower jaw of a tamarin (a) and a marmoset (b) are a means of separating these two related groups of New World monkeys. The chisel-shaped jaw of the marmosets is used to gouge holes in the bark of trees to extract gum, a major food item.

Eleven or twelve species of tamarins are recognized with more than thirty races also described. The most northerly species, the rufous-naped or Geoffroy's tamarin (*S. geoffroyi*), lives in drier forests than the more typical rainforest-dwelling Amazonian species. One of the most beautiful species, and as a result one of the most threatened from collecting for the pet trade, is the cottontop tamarin (*S. oedipus*) from Colombia. It is very strikingly marked with the long back fur dark brown, the fur on the underside pure white and the face black with a collar of rufous fur. The common name comes from the lovely white crest which runs laterally across the head from ear to ear. In northern Colombia lives a second tamarin, the white-footed (*S. leucopus*), all three species so far mentioned being closely related. In the forests of Peru, two tamarins may be found occupying the same forest habitat. They are the moustached tamarin (*S. mystax*) and the saddleback tamarin (*S. fuscicollis*), and they often move through the forest in mixed groups, though with the former species usually foraging higher up than the latter. The red-handed or midas tamarin (*S. midas*) has already been mentioned, as its range overlaps with one of the marmosets in north-eastern Brazil.

Tamarins move around the forest trees and understorey shrubs quadrupedally, with a certain amount of vertical clinging and leaping. They are principally fruit feeders, with insects coming a strong second in their diet. Like the marmosets they move around in family groups based on a breeding pair of adults with offspring of different ages.

The rare, endangered and beautiful lion tamarin, so called on account of its long mane of hair sweeping back from its bare face, encompasses three distinct subspecies all from areas of the dwindling Atlantic coast rainforest of Brazil. Apart from their striking gold or black and gold coloration they resemble the other tamarins except for their hands, which have distinctive elongated, webbed digits,. The three subspecies are the golden lion tamarin (*Leontopithecus rosalia rosalia*), which is completely gold all over, the gold and black lion tamarin (*L. r. chrysomelas*), black all over with a golden mane, and the golden-

The attractiveness of the South American emperor tamarin (*Saguinus imperator*), photographed at Twycross Zoo, perhaps goes some way to explaining why this and other tamarins are being threatened by collecting for the pet trade.

rumped lion tamarin (*L. r. chrysopygus*), black with a gold rump and legs. Some authorities, however, accept these as three distinct species – *L. rosalia*, *L. chrysomelas* and *L. chrysopygus*. Very recently a new species or subspecies of lion tamarin has been found on the island of Superagui off the coast of Brazil, about 400 km (250 miles) south-west of Rio de Janeiro. It has been named the black-faced lion tamarin (*Leontopithecus caissara*), and it is thought that no more than about two hundred individuals are present on the island. As its name implies, it has a black face; the chest, legs and tail are also black, with the rest of the body gold.

Lion tamarins live in family groups and use holes in old, hollow trees to provide them with shelter from predators at night. For this reason they rely upon mature forest to provide a safe habitat, and find it difficult though not impossible to acclimatize to the secondary forest of younger trees which is replacing much of the felled primary forest. They move quadrupedally around the forest trees, searching for their main diet of arthropods (insects, spiders, etc.) and fruit.

Finally there is a species of monkey which is classified within the Callitrichinae but uncomfortably so, for it shares characteristics with other groups as well. *Callimico goeldii* – Goeldi's monkey or Goeldi's marmoset, take your pick – is similar in appearance to the average marmoset and has their typical claws rather than nails on all but the big toes. On the other hand, it has a total of

This frontal view of the red-mantled saddleback tamarin (*Saguinus labiatus*), photographed at Twycross Zoo, shows the outward facing nostrils which are a characteristic of the New World monkeys.

thirty-six teeth in the two jaws, as against thirty-two for the other callitrichines, a feature shared with the Cebidae. The four extra teeth are a molar in each half of the upper and lower jaws. Whereas the marmosets and tamarins typically produce twins, Goeldi's monkey produces a single offspring, another characteristic shared with the cebids.

This species comes from the upper reaches of the Amazon and has a rather

The beautiful golden lion tamarin (*Leontopithecus rosalia rosalia*), photographed at Twycross Zoo, comes from the fast disappearing Atlantic coast forests of Brazil. The loss of their habitat and collecting for the pet trade has brought them close to extinction.

disjunct distribution from southern Colombia through western Brazil and into northern Peru. Although found in mature forest, it has a preference for bamboo forest and scrub where it moves by vertical clinging and leaping rather than quadrupedally like the rest of the callitrichines. It lives in small family groups and feeds upon fruit and arthropods.

Although Goeldi's monkey(*Callimico goeldii*) resembles the marmosets and tamarins in appearance, in other ways it is closer to another South American family, the Cebidae. As far as its classification is concerned it remains somewhat of an enigma. Photographed at Twycross Zoo.

THE CEBIDAE

This is the largest of the two families of New World monkeys and contains seven subfamilies with eleven genera among them. With one exception they are all diurnal.

The Cebinae

The archetypal cebids are the capuchins of the genus *Cebus*, with four species which together occupy much of the New World. The most widespread species is the tufted or black-capped capuchin (*Cebus apella*), which ranges from Colombia in the north to northern Argentina in the south and is indeed the most widespread New World monkey. The white-faced or white-throated capuchin (*Cebus capucinus*), however, is found only in Central America. Two groups of capuchins are recognized on the basis of the arrangement of the hair on the crown of the head. Tufted capuchins (*C. apellus*) have the crown covered in a dense sort of crew-cut, sometimes with tufts looking rather like horns on the sides which give the animal a somewhat pugnacious appearance. The untufted capuchins (*C. capucinus*), the white-fronted (*C. albifrons*) and the weeper (*C. nigrivittatus*) have, on the other hand, a dark crown with a peak at the front.

Capuchins are very intelligent and very successful monkeys, and they are to be found in all types of New World forest. The average head and body length for the group is roughly 450 mm (18 in); the tail, which is about the same length, is semi-prehensile at the tip, and the monkeys use it mainly as an anchor while they are at rest or feeding. They move around quadrupedally in mixed-sex groups of ten to thirty members, feeding upon whatever they can find – mainly fruit, flowers, leaves, invertebrates and even small vertebrates such as birds, if they can catch them.

The Saimirinae

The two species of squirrel monkey of the genus *Saimiri* have two completely separate distributions. The common squirrel monkey (*S. sciureus*) is found throughout most of the South American rainforest, whereas the red-backed squirrel monkey (*S. oerstedii*) is restricted to Costa Rica and Panama. They are the smallest of the cebids, with head and body together 320 mm (12.5 in) and the tail one-third as long again; males are distinguishable by their well-developed canine teeth. They move through the forests quadrupedally, troop size depending to a great extent upon the area of forest available to them. In the Amazon forests groups of as many as five hundred individuals have been reported but in smaller areas numbers average around twenty-five per group. Within each group males, mothers with offspring and juveniles tend to form subgroups, though this system breaks down in the breeding season. Their main food is fruit, with an important supplement of arthropods such as insects and spiders.

The somewhat brutish appearance of the brown capuchin monkey (*Capucinus albifrons*), along with the other capuchins, belies the fact that they are the most intelligent and resourceful of the South American monkeys. Photographed at Twycross Zoo.

The Callicebinae

The members of this subfamily are referred to as titis and there are three species, two from the Amazon forests and a third from the Brazilian coastal rainforest. They are slightly larger than the squirrel monkeys and have a much bushier coat. The widow or white-handed titi (*Callicebus torquatus*) and the dusky titi (*C. moloch*) live together in the same areas of forest but occupy

Squirrel monkeys are very efficient foragers, their troops leaving nothing undisturbed in their search for fruit, flowers, leaves and insects as they move through the forest trees. This is the red-backed squirrel monkey (*Saimiri oerstedii*) in rainforest in Costa Rica.

different ecological niches. Dusky titis prefer the wetter areas of the forest, whereas widow titis are restricted to the drier areas. Fruit forms the main food of both species, but the widow titi also eats insects whereas the dusky titi supplements its diet with leaves. The third species is the masked titi (*C. personatus*), and in the absence of any competing species it is able to make broader use of its environment than the others. All three species are in the main quadrupedal, with some vertical clinging and leaping, and they move around their forest domain in small family groups of a male and female pair with their offspring.

The Alouattinae

This subfamily contains the noisiest of the New World monkeys, the howlers, of which there are six species ranging over a huge area from the rainforest of southern Mexico down into northern Argentina. Their loud calls are produced from an egg-shaped structure below the chin, which acts as a resonating chamber. Male howlers produce the deepest calls, and accordingly have the largest resonating chambers. They are large monkeys, twice the size of the squirrel monkeys, and they possess a prehensile tail. All species show sexual dimorphism in relation to body size, the females being only about three-quarters the size of the males. In one species, the black howler (*Alouatta caraya*), there is marked sexual dichromatism: the adult males are black while the females and young are a pale creamy grey. The species, too, are differentiated on the basis of their coloration. The Guatemalan howler (*A. villosa*) is completely black, while the other Central American species, the mantled howler (*A. palliata*), is also black but with a gold fringe along the flanks. The third black species, the black and red howler (*A. belzebul*), has red hands, feet and tail-tip; the brown howler (*A. fuscas*), from eastern Brazil, is dark brown; and finally the red howler (*A. seniculus*), from Venezuela and Trinidad, is coppery red.

The howlers move quadrupedally in the middle to upper storeys of the forests, seldom rushing, and using their prehensile tails as an anchor when they are feeding or at rest. They live in multi-male groups ranging in size from just three to as many as twenty-five individuals, and feed exclusively on plant material. Some feed mainly on leaves, while others eat fruit with a supplement of leaves. Despite being vegetarians their gut shows no special modifications for the digestion of large quantities of plant material.

The Atelinae

This subfamily includes the spider monkeys and the woolly monkeys, which are found in three separate genera. Of the three the spider monkeys of the genus *Ateles* are the most widespread, with four species extending from Mexico in the north down as far as northern Argentina, and they are separated by some taxonomists into a number of distinctive races. They are similar in size to the howler monkeys, but their limbs are more noticeably long and slender. Of the four species two – the black spider monkey (*A. paniscus*) and the brown-headed spider monkey (*A. fusciceps*) – are basically black in colour, although as its name implies the latter sometimes has a brown crown. The long-haired spider monkey (*A. belzebuth*) from Colombia, on the other hand, is black or brown with pale underparts and forehead, while the black-handed or Geoffroy's spider monkey (*A. geoffroyi*) varies from gold through red to buff or dark brown, with various black markings. The sexes are similar in size, but the males have markedly longer canine teeth than the females.

Spider monkeys are highly arboreal and are found in a variety of forest types including mangroves. They move around quadrupedally, but also use brachiation, and if need be can leap considerable distances into dense masses of flexible foliage to break their fall. The tail is highly prehensile and easily supports the animal's whole weight. All species rely heavily upon ripe fruit as a food source, supplemented with young leaves. They move around in single to multi-male groups of from six to roughly thirty individuals, with the larger groups divided up into smaller subgroups for much of the time.

In the Atlantic rainforest of Brazil *Ateles'* niche is taken by the now rare muriqui (*Brachyteles arachnoides*). It was formerly called the woolly spider monkey, since it resembles an *Ateles* in shape but has the more rounded head and short, dense fur of a woolly monkey. It is, however, considerably larger

The howler monkeys range from southern Mexico, down through Central America and over most of South America. These are females of the sexually dichromatic black howler (*Alouatta caraya*) in Twycross Zoo; the male is all black. The sad faces are not because they are unhappy – they even look like this in the wild.

than *Ateles*; the head and body length for *Brachyteles* is 780 mm (30 in) as against an average of 560 mm (22 in) for the former. The woolly spider is in fact the largest of the New World monkeys. It moves around the forest canopy in groups in much the same way as *Ateles*, its diet consisting of fruit, seeds and leaves.

The third ateline genus, *Lagothrix*, contains the two species of woolly monkeys. The common woolly monkey (*L. lagotricha*) is widely distributed in the upper Amazon forests from Colombia through Brazil into northern Bolivia, while the yellow-tailed woolly monkey (*L. flavicauda*) is very rare and is confined to the dense cloud forests around 1800 m (6000 ft) up in the Peruvian Andes. They are large monkeys, with a head and body length of around 520 mm (20.5 in), and have markedly snub-nosed faces, dense short fur and a long prehensile tail. The male canines are longer than those of the female; but the most odd thing about them is that the female clitoris is longer than the male penis, making it difficult to distinguish the sexes in the wild. They feed mainly upon fruit, with a supplement of seeds and leaves. They move quadrupedally, but with some brachiation, in mixed sex groups of around twelve individuals.

The Pitheciinae

Contained in this subfamily are the sakis of the genus *Pithecia* and *Chiropotes* and the uakaris of the genus *Cacajao*. The two species of *Pithecia* are medium-sized monkeys with long, shaggy coats; the hair along their forehead forms a sort of hood and the long, thick fur continues on to the non-prehensile tail, making it look markedly fat. The nose is so broad that the nostrils are not visible from the front. Both sexes of the monk saki (*P. monachus*) are similar, with dark brown to black fur tipped with white and a line of pale fur running from below the eyes and down each side of the face. The white-faced saki (*P. pithecia*), however, shows marked sexual dichromatism. The male is completely black with an off-white face, while the female resembles the previous species except for its reddish underparts.

The two pithecias are found over much of the Amazon basin as far as the Guianas in the north and Bolivia, Colombia, Ecuador and Peru, in the west, but they have a limited distribution southwards in Brazil. They are tolerant of most types of forest, but shy away from those that are subject to flooding; they prefer the middle layers of the forest where they feed upon fruits and seeds with the occasional young shoots and leaves. Pithecias live in small family groups and progress through the forest quadrupedally, though they also have the ability to make quite long downward leaps into suitably flexible foliage.

In contrast to the long, shaggy fur of the sakis, that of the two species of bearded saki is short and thick. They are most easily distinguished by their beards and by the bulging forehead of the males. The black-bearded saki (*Chiropotes satanas*) is found in the north-east of the Amazon basin, whereas the red-nosed bearded saki (*C. albinasus*) is found south of the Amazon and north of the Madeira River. Despite the error in the original Latin name, which inferred that it had a white nose, *C. albinasus* is distinguished from *C. satanas* by its red face. They are roughly equal in size to the sakis and move around in much the same fashion, though they have been observed to leap less often. They spend much time searching out trees with the best seeds and fruit, which form the major part of their diet. Bearded sakis have forward-projecting incisors

The white-faced saki (*Pithecia pithecia*) is found over much of the northern Amazon basin area. It is one of a number of primates which exhibit sexual dichromatism. This is the male; the female is brown to black all over, with white tips to the hairs and no white face mask. Photographed at Twycross Zoo.

which are used to break open the hard shells of some fruits and nuts so that they can feed on the soft kernels. Foraging takes place in mixed-sex groups of thirty or so individuals.

Pride of place for ugliness amongst the New World monkeys must surely go to the three species of uakari, whose distribution is restricted to the upper Amazon basin. They are medium-sized, short-tailed monkeys with long, shaggy coats, and two of the three have bright red, naked faces and bald heads. The bald or white uakari (*Cacajao calvus*) has white or yellowish hair, while the red uakari (*C. rubicundus*) has rust-coloured fur; both have a red face. Some authorities consider these two to be races of *C. calvus*. The third species, the black-headed uakari (*C. melanocephalus*), is somewhat less shaggy, with yellowish brown fur and red legs and tail. As its name implies, it is not bald like the others but has black hair on the head and also on the arms, hands and feet.

The uakaris appear to be restricted to the Amazonian inundation forests, i.e. regions of forest flooded during the wet season, where they are found at all levels; and, like the bearded sakis, they are specialist seed and nut feeders. They move around quadrupedally and also have considerable leaping abilities. Both species form mixed-sex groups of between fifteen and thirty individuals.

The Aotinae

This subfamily contains just a single species, the douroucouli, also known as the night or owl monkey (*Aotus trivirgatus*); it is the world's only nocturnal monkey and the only nocturnal primate from the New World. It has a very wide distribution, from Panama in the north to the Gran Chaco of Argentina and Paraguay in the south, though it is not found in the Atlantic coast forests of Brazil. The large eyes and nocturnal habits prevent confusion with any other species. They feed at all levels in the forest, mainly on fruits, but they also take young shoots, leaves and flowers and a range of small animal prey if they come across it. Their locomotion is mainly quadrupedal, though they can also leap from tree to tree, even in the dark. Family groups are the norm.

Chapter 3

The Old World Monkeys and the Apes

THE CERCOPITHECIDAE

This family is divided into two subfamilies and contains all the Old World monkeys. The subfamilies consist of the mainly omnivorous Cercopithecinae, which includes the guenons, talapoin and baboons from Africa, and the principally Asian macaques and the vegetarian Colobinae, including the African colobus monkeys and the Asian langurs.

The Cercopithecinae

The cercopithecines include some of the most adaptable and intelligent of all non-hominoid primates. Monkeys of the genus *Cercopithecus*, collectively called guenons, are very widespread across the whole of Africa south of the Sahara with the exception of the eastern Cape region. There are roughly twenty species and a very much greater number of subspecies and races.

Guenons have the following features in common. They are medium to large monkeys with head and body between about 500 mm (19 in) and 570 mm (22 in) in length, and a tail considerably longer than this. Males are generally somewhat bigger and quite a lot heavier than females. The head is rounded and the muzzle rather short, with pronounced cheek whiskers standing out sideways from the face; in addition, some species have a beard. They all have well-developed cheek pouches, enabling them to feed rapidly in an exposed position and then to retire to safety to chew and swallow their food at their leisure. Body hair colour and marking are quite variable and one or two species, for example the Diana monkey (*C. diana*), are very attractive. The species are differentiated mainly on the nature of their face patterns, but most of them can also be distinguished from behind as well. This is important for the monkeys, for in some areas of Africa a number of species live sympatrically and so they need to be able to recognize their own. Where species live together in the same environment they tend to forage at different levels or in different ways, so that competition between them is reduced to a minimum.

There is such a plethora of common names for the individual species and subspecies that the situation becomes quite confusing. Collectively they are referred to as guenons, but within the group as a whole there are the Diana, Mona, green, vervet, grivet and tantalus monkeys, all reflecting different groupings or subspecies within these groups. As often as not, however, common names denote some outstanding feature of a species or subspecies – redtail monkey, for example, is self-evident. Complications arise when certain sympatric species, from time to time, hybridize one with another.

The Diana monkey(*Cercopithecus diana*) is found mainly in West Africa and has a number of subspecies. It is one of the more handsome of the cercopithecines; this individual is resident in Twycross Zoo.

The majority of species live an arboreal life in the varied forest types found throughout their range. The species with the greatest range and also the largest number of subspecies (twenty-one are recognised by some authorities) is, however, partly adapted to ground living. The green monkey (*C. aethiops*) and all its forms, such as the grivet, the vervet and the tantalus, are all savannah and steppe dwellers. They spend the night in the safety of the trees, and during the day forage both in the trees and on the ground. They have also

The various species of cercopithecine monkeys are distinguished from one another partly on the basis of their facial adornments. This attractive little face is of a female Schmidt's guenon (*Cercopithecus ascanius schmidti*), photographed at Twycross Zoo.

taken to raiding cultivated areas – so much so that they have become a pest in some places. One species which is something of an exception to the general diurnal, omnivorous *Cercopithecus* is the so-called owlfaced monkey (*C. hamlyni*) from the eastern Zaire basin. Not only is it vegetarian, feeding mainly on bamboo shoots, but it is also said to be nocturnal.

Perhaps the rarest of the guenons is *C. sclateri*, originally *C. erythrotis sclateri*, Sclater's russet-eared guenon. This animal was first described in 1904 from an animal at London Zoo, which died not long after its arrival and its remains were then deposited in the British Museum (Natural History). Its precise origins were unknown, apart from the fact that it came from somewhere in what is now Nigeria. A search was made for it in 1982 in the area from which it was thought to come, but with negative results. In January 1988, however, an expedition to the Niger flood plain was successful and the monkey was seen in the wild for the first time.

All the *Cercopithecus* species move around quadrupedally, though they also have considerable leaping powers and at all times use their long tails to help maintain their balance. In the tree-dwelling species, groups are based on a single male; but in the essentially ground-dwelling green monkeys more than one male may be found per group, with one dominant over the remainder.

With its own subgenus is Allen's swamp monkey or the blackish green guenon (*Allenopithecus nigroviridis*). This unusual monkey is somewhat macaque-like in appearance and inhabits the swamp forests of the central Zaire basin. In the past its habitat has been difficult to access, and little is known about its behaviour except that it forages in groups feeding on fruits, leaves and other vegetable material as well as snails and insects, and that it also descends from the trees to catch freshwater crabs and small fish.

The Old World's smallest monkey, the talapoin (*Miopithecus talapoin*), has its widest distribution in the lowland forests of Cameroun and Gabon but it is also found in similar situations in Lower Congo and Zaire, with a further disjunct population to the south in Angola. Given these two fairly widely separated main populations some authorities consider there are distinct northern and southern species, while others break them down into four subspecies.

The lifestyle of the talapoins is in many ways similar to that of the South American squirrel monkeys, which they resemble in size. They are very omnivorous and will eat almost anything edible that they can get into their mouths. One aspect of their lives which is unusual is their excellent swimming ability: if danger threatens they will enter water at the earliest opportunity and dive to a considerable depth in order to escape; if possible they spend the night on trees overhanging water, in order to ensure that this escape route is available. They live in mixed-sex groups of up to 150 individuals, within which it is possible to find subgroups comprising all males, females with young, or juveniles with an adult male 'babysitter' – a further parallel with the New World squirrel monkeys.

At the opposite extreme to the tiny talapoin is one of the larger species, the patas, military or hussar monkey (*Erythrocebus patas*). Although it sleeps in trees at night, the rest of the time it is almost entirely terrestrial. Its most obvious

Allen's swamp monkey (*Allenopithecus nigroviridis*) is an inhabitant of the swamp forests of the central Zaire basin in Africa. Here, a female is grooming her half-grown offspring in Twycross Zoo.

features are its very long arms and legs, with which it can produce a considerable turn of speed in escaping from its main enemies, cheetahs, leopards, hyenas and hunting dogs. Running speeds of as high as 55 kph (35 mph) have been recorded for patas monkeys. They are inhabitants of savannah and open grassland, sometimes straying into cultivated areas, in sub-Saharan Africa west of the Nile but north of the Zaire basin. A number of subspecies are recognized, based mainly upon facial features and distribution.

Patas monkeys live in groups of up to thirty individuals, though twenty is average, with a single old dominant male whose major role is to act as a lookout and who, if the group is threatened, will create some form of diversion

65

while the rest make good their escape. He is accompanied by a number of females and their offspring, the young males remaining single or in small all-male groups.

The next group of cercopithecines, the mangabeys of the genus *Cercocebus*, have longer muzzles than the preceding genera and are closely related to the baboons. In fact it has been said that the grey-cheeked mangabey (*C. albigena*) is probably more closely related to baboons of the genus *Papio* than it is to the other mangabeys. They are found in a range of forest habitats, including mangrove and swamp forest in the Zaire basin area and in West Africa from Guinea to Nigeria. Two species are mainly arboreal, while the other two also forage on the forest floor.

The four species are fairly easily distinguishable, with *C. galeritus*, variously named the Tana River, crested, agile or golden-bellied mangabey, the most obvious with its brown to rufous coloration. The basic coat colour of the other three species is dark grey to black. They are fairly large monkeys, with a head and body length up to 730 mm (28 in) for some male individuals of *C. albigena*, though the other species are a few inches shorter and females are somewhat smaller than males. Their food consists mainly of fruit and leaves, supplemented with arthropods and some fungi which they can store temporarily in their large cheek pouches. They move around quadrupedally in single or multi-male groups.

The remainder of the cercopithecines have noticeably long muzzles and include the baboons, the mandrill, the drill and the gelada. The taxonomy of these animals is in some disorder: some authorities place them all in the genus *Papio*, while others put the drill and the mandrill into *Mandrillus* and the gelada into *Theropithecus*. For convenience we will use the latter classification here.

The genus *Papio*, the baboons proper, contains either two species, one of which contains four subspecies, or five distinct species – again depending upon the taxonomic authority consulted. They extend over much of Africa south of the Sahara, with the exception of certain of the heavily forested areas, and the hamadryas or sacred baboon (*P. hamadryas*) also extends into south-west Arabia. *Papio* are principally inhabitants of wooded savannah, grassland, acacia scrub and semi-desert as long as they have access to water. In places where open water is not freely available they may actually dig down to find it.

They are large, dog-like, terrestrial animals with sexual dimorphism in size, the males being about twice the weight of the females. The males have a more massively built skull than the females, with a sagittal crest, and the male upper canine teeth are massive. As well as size dimorphism the hamadryas baboon also demonstrates sexual dichromatism: the male is clothed handsomely in silvery grey with a long cape over the shoulders and long, white sideburns, while the female is just olive-brown. Of the others the olive baboon (*P. anubis*) is olive-brown, the yellow baboon (*P. cynocephalus*) is yellow-brown to grey, the Guinea baboon (*P. papio*) is reddish brown and the chacma baboon (*P. ursinus*) is blackish brown, males and females being similarly coloured.

Baboons move around quadrupedally in bands of up to two hundred, with the dominant males leading and the lesser males bringing up the rear, females

The elongated muzzle of this male black-crested mangabey (*Cercocebus aterrimus*), which comes from Africa's Zaire basin region, is an indication that these monkeys are closely related to the baboons. Photographed at Twycross Zoo.

and youngsters in between them. The exception is the hamadryas baboon, which spends the night in large aggregations of several hundred individuals on rock outcrops and cliffs to be safe from predators. During the day they disperse and forage in groups of a single male accompanied by his harem of up to twelve adult females. Baboons are all highly omnivorous and will eat almost anything that is edible, including small mammals if they chance upon them. They also dig up roots and tubers, and those that live near the sea forage along the shore for crustaceans and other edible marine life.

Whereas the baboons are only occasionally forest dwellers, this way of life is

This female hamadryas baboon (*Papio hamadryas*), in the company of two youngsters of different ages at Munich Zoo, shows the typical dog-like appearance of the baboons as a group.

the norm for the mandrill (*Mandrillus sphinx*) and drill (*M. leucophaeus*). Both are found in western central Africa, from south-eastern Nigeria in the north; the mandrill reaches as far south as the Zaire River, the drill not quite so far south. Like the baboons, the males of these two species are considerably heavier than the females and they show some sexual dimorphism in the shape and size of the muzzle. That of the males is more massive, and there are bony swellings along the side of the nose. In the male drill the face and muzzle are black, with transverse red bands on the lower lip. The male mandrill, on the other hand, has one of the most incredible of all faces – the front and top of the nose are red, the sides of the nose are blue and the chin and side whiskers orange. What is more, the rear ends of the males of both species are coloured red, blue and violet. Unlike the medium-length tail of the baboon and the gelada, that of the drill and the mandrill is just a short stump.

The heavy males of both species are mainly terrestrial, but the females and their offspring climb into the forest trees to feed. Like the baboons they move quadrupedally when foraging, but are less omnivorous, the emphasis in their diet being on fruit, leaves and ground vegetation with some insects. As forest dwellers they are not as easy to observe as the baboons, so less is known about them, but it is believed that they form one- or multi-male groups which may coalesce to form troops of up to 150 individuals.

The gelada (*Theropithecus gelada*) has a limited distribution on mountains above 2000 m (6500 ft) in central Ethiopia. The animals retreat at night to cliffs along the steep gorges that are typical of these high plateaus, emerging in the day to feed on the adjacent montane grasslands. They are seldom found far from their cliffs, and retreat to them whenever danger threatens. In appearance they resemble the hamadryas baboon, with which they sometimes mix where their ranges overlap, but they are greyish to dark brown in colour and the male has a sweeping blackish brown cape of long hairs and long, whitish sideburns. Both sexes have patches of bright pink skin on the front of the neck and extending down on to the chest.

Like the baboons, the geladas are quadrupedal, but they also spend a good deal of time shuffling around on their haunches grazing, using their highly opposable thumb and index finger to pluck off plant material. To keep them comfortable they have a pair of fatty pads on their rear ends, similar to human buttocks. They are almost exclusively vegetarians, with a main diet of grass and grass seeds supplemented by fruits and other plant material if available, and a few insects. The basic foraging groups are single male-dominated harems which come together into much larger groups of several hundred individuals on the sleeping cliffs.

The macaques of the genus *Macaca* form a group that is restricted to Asia, from India eastwards to Japan, the Philippines and Timor, with the exception of a single species found in North Africa. They are usefully subdivided into four well-founded species groups based upon the structure of the male and female reproductive systems and copulatory mechanisms. These subdivisions have also been found to be well supported by serological investigations.

The macaques are sturdily built monkeys of medium size, with the tail varying in length from none at all to longer than the head and body length. They are found in a very wide range of habitats in forests of all kinds, including mangroves, and a number of species are well adapted to living in close proximity to humans. They have a noticeable but not long muzzle and well-developed cheek pouches, and in accumulating their omnivorous diet they are assisted by having a fully opposable thumb. Males have well-developed canines and are considerably bigger and heavier than females, and as with the baboons the upper canine is honed by the first lower premolar. Most species are greyish to brownish in colour. They move quadrupedally and, while most species are primarily terrestrial, a few are highly arboreal. Socially they form multi-male groups with a dominance hierarchy in both the male and female members of the group.

The first of the four groups is based upon the crab-eating or long-tailed macaque (*M. fascicularis*), which is found in the south-east corner of Bangladesh and extends from there as far east as the Philippines and Timor. The Taiwan or Formosan rock macaque (*M. cyclopis*) is found only on the island of Taiwan. At one time it often lived in close association with humans, but due to collecting for research and hunting it now restricts itself to forest habitats. The third member of the group is the Japanese macaque (*M. fuscata*) from all Japan except the island of Hokkaido. Finally in this group is the rhesus macaque (*M. mulatta*), which is found from eastern Afghanistan through Pakistan and India into China. In India they have developed a very close association with humans, but otherwise their distribution is dictated by habitat.

The second group is the 'silenus-sylvanus' group, based upon the lion-tailed macaque (*M. silenus*) and the barbary macaque (*M. sylvanus*). *M. silenus* is restricted to the fast disappearing deciduous forests of the Western Ghats in the south of India. It is a magnificent black creature with a lion-like mane of contrasting grey hair around the face. Their diet is much more restricted than most macaques for they feed almost entirely upon fruit. The barbary macaque is today found in the temperate mountain forests of Algeria and Morocco and it has also been introduced into Gibraltar, though in prehistoric times it occurred naturally in south-western Europe. Like most of the macaques the barbary is a medium-sized, sturdily built monkey, but it is one of the only two species which lacks a tail. It sleeps in trees or amongst rocks but is otherwise terrestrial. Mixed-sex groups of between ten and thirty individuals forage for plant foods such as grass, fruits, young leaves, bulbs and tubers as well as arthropods such as insects and spiders. The pig-tailed macaque (*M. nemestrina*) is found over much the same range as *M. fascicularis* but they occupy different habitats – the former tends to be encountered less often than the latter, since it is mainly an inhabitant of primary rainforest.

Also included in the 'silenus-sylvanus' group are the seven species of Sulawesi macaques. Not a great deal is known about them though they are said to have non-overlapping distributions on the island, where they reach their greatest numbers in the densely forested areas. The species concerned are the Sulawesi crested macaque (*M. nigra*), which is entirely black, the Sulawesi booted macaque (*M. ochreata*), the Moor macaque (*M. maura*), the Tonkean macaque (*M. tonkeana*) and three without common names: *M. brunnescens*, *M. hecki* and *M. nigrescens*.

The 'sinica' group is based upon the toque macaque (*M. sinica*), which lives in a range of forest types in Sri Lanka. The Assamese macaque (*M. assamensis*) inhabits forests in the foothills of the Himalayas and neighbouring mountains in south-east Asia. This species is known to be a climber and spends much of its life high up in the forest trees, where fruit forms a major part of its diet. The bonnet macaque (*M. radiata*) lives across much of southern India in various forest types where, like the previous species, it spends much of its life in the trees; but it reaches its greatest numbers in cultivated areas close to towns and villages. In the evergreen forests of this area they are sympatric with the lion-tailed macaque. The bonnet macaque shares with the toque the feature of having a tail longer than the head and body combined. The other member of the group is the Tibetan macaque (*M. thibetana*) from the mountains of east central China, where it inhabits forest remnants of all types. This is one of the terrestrial macaques and, although it is mainly vegetarian, it will also take small animals and birds' eggs.

The lion-tailed macaque (*Macaca silenus*) is an endangered species since it comes from the fast-disappearing deciduous forests of the Western Ghats of southern India.

The fourth group contains a single species – the bear macaque (*M. arctoides*), a rather ugly, dark brown creature with a bare red face and forehead. It is found from southern China across to eastern India and down into the north of the Malaysian peninsula. These animals are forest dwellers and, though mainly vegetarian, they supplement this diet with small animals; where their habitat has been destroyed by humans they often hang on and feed upon the planted crops.

The Colobinae
Whereas the cercopithecines are generally highly omnivorous, the colobines are wholly vegetarian. Related to this the colobines lack cheek pouches and their stomachs are large and divided into chambers. A third factor which distinguishes them from the cercopithecines is that their eyes are generally placed further apart. The two largest genera include the African colobus monkeys and the Asian langurs.

There are six species within the genus *Colobus*, four of which come under the general heading of black and white colobus while the other two are red colobus. The black and white colobus are fairly large monkeys: males of some species attain a head and body length of 700 mm (28 in), with the females somewhat shorter. Their tails are noticeably long, up to $1\frac{1}{3}$ times the head and body length. The western black and white colobus (*C. polykomos*) has one series of subspecies ranging from Guinea in the west to western Nigeria, and then a much larger area of population with a further series of subspecies in Cameroun, Gabon, Congo (Brazzaville), Zaire, northern Angola and eastwards into western Uganda and Tanzania. Some authorities raise some of these subspecies to the level of species: for example, *C. satanas* from Cameroun, Gabon and part of Congo (Brazzaville) is all black, while the mantled or Angolan colobus has very long side whiskers and beard forming a crescent shape. The guereza or eastern black and white colobus (*C. guereza*) ranges from eastern Nigeria eastwards, but north of the previously described species, through Chad, the Central African Republic and southern Sudan into Kenya and Tanzania and north into Ethiopia.

The red colobus (*C. badius*) is slightly smaller than the black and white group, while the olive colobus (*C. verus*) is the smallest of all at only 450 mm (17 in) head and body length. The subspecies of red colobus show a fair degree of colour variation, but in combinations of brown, reddish brown to orange and black with pale undersides. They have a rather disjunct range: for example, *Colobus badius temmincki* is found from The Gambia into north-western Guinea, and then there is a gap to *C. b. badius*, which extends from southern Sierra Leone through Liberia and into the Ivory Coast where *C. b. waldroni* takes over, reaching as far as western Ghana. The largest single distribution covering a number of subspecies is in central Africa around the Zaire basin into western Uganda and Tanzania. A totally isolated population of *C. b. rufomitratus* occurs in the coastal forest region of south-eastern Kenya. On its own is the protected but still threatened Kirk's colobus (*C. b. kirkii*) from the Jozani Forest on the island of Zanzibar. Finally comes *C. verus*, the olive colobus (sometimes placed in *Procolobus*); it is sympatric with *C. b. temmincki*, having almost the same range.

Colobus monkeys move around quadrupedally but they also resort to semibrachiation at times. They spend almost all of their lives in the trees, seldom descending to the ground. They feed upon fruit, leaves of all ages, flowers and even twigs, with the red colobus types foraging on a greater range of tree species than the black and white forms. Foraging groups are based upon a single male with a number of females and their offspring, and these may unite at times to form larger groupings.

The remainder of the colobine monkeys all come from Asia; the largest genus is *Presbytis*, the leaf monkeys or langurs. They are found over the whole of the Indian subcontinent across into south-western China and then down the whole of Indo China and Malaysia and on the islands as far east as Borneo and Sumbawa in the Greater Sunda Islands. With the exception of the grey or Hanuman langur (*P. entellus*), which is mainly terrestrial, they are all highly arboreal, feeding upon all edible parts of the trees they inhabit. They are medium to large monkeys with long tails and long hind-limbs, an indication of the fact that they are good leapers, though they also move quadrupedally and

The guereza or eastern black and white colobus monkey (*Colobus guereza*) has a range extending across Africa from Nigeria in the west to Kenya and Tanzania in the east, and north into Ethiopia. Photographed at Twycross Zoo.

by semi-brachiation. There is some degree of sexual dimorphism in size and weight in the larger species, but this is less marked in the smaller species. The langurs form groups with one or more males in the company of the females and young, with the exception of the Mentawai langur, which forms family groups. As with so many monkeys, spare males often group together and forage.

Langurs fall into four distinct groups, based most interestingly upon the fur colour of the newborn young, which differs markedly from that of their parents. The Hanuman langur is the only one in its group and with its long association with humans in India as a sacred animal it is the species about which most is known. The males can become very large indeed, weighing over

18 kg (40 lb). Adults are pale grey, while in contrast the newborn young are dark grey. In the group headed by the silvered langur (*P. cristata*) and containing five other species with grey to black adults, the young have bright apricot fur. The third group, the banded langur (*P. melalophos*) group, contains adults which are black, grey, reddish or brown, but all of the young are white with a dark stripe down the back as far as the tip of the tail. The final group contains two species in which the adults are of the same colour, grey-brown to black, but they have different-coloured young. Those of the purple-faced langur (*P. vetulus*) from Sri Lanka are pale grey, while the young of the Nilgiri langur (*P. johnii*) from south-west India are red-brown.

Langurs live sympatrically with other langurs in some areas, and with cercopithecines in others. For example, in the Malaysian peninsula banded lan-

A Temminck's red colobus monkey (*Colobus badius temmincki*) photographed in gallery forest along the Gambia River. This picture clearly shows the right-hand ischial callosity on the monkey's rear end.

gurs (*P. melalophos*) and spectacled langurs (*P. obscura*) are found together, but competition is reduced by their different foraging habits. Much the same is true for maroon langurs (*P. rubicunda*) and white-fronted langurs (*P. frontata*) in parts of Borneo. In the dry deciduous forests of southern India the Hanuman langur is sympatric with the bonnet macaque.

The remaining Asian colobines are from four different genera, each one

The spectacled langur (*Presbytis obscura*) from Malaysia is one of a number of species, sometimes called leaf monkeys, which are found across Asia from the Indian subcontinent in the west to the Greater Sunda Islands in the east and north into south-western China. Photographed at Twycross Zoo.

very distinctive in appearance. There are few monkeys to which the adjective 'beautiful' can be applied, but this is surely true of the douc (*Pygathrix nemaeus*) from the forests of Laos and Vietnam. The douc is a fairly large monkey with the tail just longer than the head and body; males and females are of similar size. Two subspecies are recognized: the douc (*P. n. nemaeus*) has a pale face and black thighs, with the rest of the legs red and white, and cuffs on the forearms; while the black-footed douc (*P. n. nigripes*) from southern Vietnam has all-black legs and a black face, and lacks the white cuffs. The main observable difference between the sexes is that the white hairs which encircle the face are much longer in males. The douc is primarily arboreal, where it moves quadru-pedally but also leaps a great deal. A typical colobine, it is a vegetarian, feeding on young leaves, fruit and flowers. The doucs form groups of one or more males with two or more females, up to a maximum of about twelve indi-viduals.

A second monotypic genus contains the pig-tailed leaf monkey or simakobu (*Simias concolor*) from the Mentawai Islands west of Sumatra; it is also referred to as the Mentawai Islands snub-nosed monkey. These stocky, medium-sized monkeys are the only colobine with a short tail – just one-third of the head and body length, and almost hairless. Simakobu have a pronounced snub nose and black faces, and they exist in two colour forms within the population so that about two-thirds are black and the rest cream. They move around quadru-pedally in their tropical rainforest habitat, where they feed upon leaves and fruit. The normal grouping is based upon one male, one female and their offspring.

Closely related to the previous species and perhaps sharing a common ancestor is the bigger and better-known proboscis monkey (*Nasalis larvatus*), whose range is restricted to the mangrove forests of the island of Borneo. They are large monkeys: the male reaches around 750 mm (29 in), but with the tail a little shorter than this, and tops the scales at 20 kg (44 lb), with the females roughly half this weight. The proboscis monkey's most obvious feature is the large pendulous nose, which in males hangs down to below the level of the mouth, though in the female it is considerably shorter. Proboscis monkeys spend most of their lives up in the mangroves where they feed mainly upon young shoots and leaves with a supplement of fruit and flowers. They move quadrupedally through the trees, but where one area is separated from another by a deep-water channel they are quite prepared to swim across. If danger threatens, they are also able to dive and swim under water. They form multi-male groups of up to sixty individuals, though the average group size is around twenty.

Finally we come to the snub-nosed monkeys of the genus *Rhinopithecus* – large, heavily built animals whose females weigh up to 10 kg (22 lb) and whose males are one-third as heavy again. The skin of the muzzle and around the eyes is pale blue, and they have markedly upturned noses. The Chinese snub-nosed monkey, golden monkey or snow monkey (*R. roxellana*) comes, as its name implies, from the high mountains in the west of China, where it has to endure very cold winters and not particularly warm summers. As a result they

The beautiful douc langur is most likely to be seen in captivity, since it comes from the inaccessible forests of Laos and Vietnam. This is the northern subspecies *Pygathrix nemaeus nemaeus*.

change their coats each autumn. A second, smaller, darker species with paler underparts, the Tonkin snub-nosed monkey (*R. avunculus*), is known from the tropical forest of northern Vietnam, though there is little information on its present status.

Our knowledge of the ecology of these monkeys is somewhat limited. They

Figure 9 The male of the proboscis monkey (*Nasalis larvatus*), from the mangrove swamps of Borneo, has a very enlarged nose, though no one knows quite what its significance is.

are known to move quadrupedally and can also leap from tree to tree; the golden monkey is also said to move from the higher slopes into the valleys in winter, indicating some movement across open ground when necessary. During the winter months males and females spend a lot of time hugging each other, nuzzling their faces into each other's fur to keep warm.

THE HYLOBATIDAE

This family contains a single genus, *Hylobates*, the gibbons and the siamang. Head and body length of both sexes in the gibbons is about 440 mm (17 2 in) and they weigh 5–7 kg (11–15 lb). The siamang is a larger animal up to 540 mm (21 in) in length and weighing up to around 27 lb. They are highly arboreal, inhabiting primary forests from Assam across to China and south into south-east Asia, peninsular Malaysia and the islands of Borneo, Java and Sumatra as well as two of the Mentawai Islands off Sumatra. Despite their tropical distribution they have dense, fluffy fur. Some species show sexual dichromatism. In the black gibbon (*H. concolor*), the hoolock gibbon (*H. hoolocki*) and the pileated gibbon (*H. pileatus*) the males are all-black, whereas the females are brown, grey, buff or golden. Both males and females can be either black or buff in the lar gibbon (*H. lar*), while both sexes of the Mentawai Island gibbon (*H. klossii*) are black and both sexes of the siamang

A number of gibbon species come from both the mainland and islands of south-east Asia. This lar gibbon (*Hylobates lar*) in Metro Zoo, Miami, is hanging by the long arms characteristic of these apes, which move through the trees by brachiation. Both males and females of this species can be buff, as here, or all-black.

(*H. syndactylus*) have long black fur. Gibbons have highly melodious voices, and the siamang in particular produces two distinct notes from an air sac lying below its chin.

The gibbons and siamang have long arms and relatively short legs, and among the apes they are the only true brachiators. All species of *Hylobates* are basically vegetarian: the gibbons subsist primarily upon fruit and supplement this diet with leaves, flowers and some insects, while the siamang eats roughly equal quantities of fruits and leaves, again with a supplement of flowers and insects.

THE PONGIDAE

Of the so-called great apes the orang-utan (*Pongo pygmaeus*) is the only Asian representative. Although in prehistoric times there were orang-utans on the Asian mainland, today they are found in just a few restricted areas of the islands of Borneo and Sumatra. Each island has its own subspecies and these are easily recognizable by the shape of the face. The Bornean subspecies *P. p. pygmaeus* has a noticeably round face, whereas that of the Sumatran subspecies *P. p. abelii* is long and narrow and it also has paler, longer fur. Male orangs with recorded weights of between 45 and 100 kg (100–220 lb) are considerably larger than females, whose weight ranges from 35 to 50 kg (75–110 lb).

Orangs are essentially arboreal and move slowly through the trees, grasping branches with both their hands and feet. They feed mainly upon fruit, with a supplement of insects; when fruit is not available they fall back on leaves and tree bark. At night they make substantial foliage nests in the trees, for they lack ischial callosities and would therefore find it uncomfortable to sleep in a sitting position. Groups of orangs are likely to be mothers with their offspring of various ages; the males only take an interest in the females when they are sexually receptive.

For the remaining pongids, the gorilla and the chimpanzees, we have to return to Africa. There is a single species of gorilla (*Gorilla gorilla*), with three distinct subspecies. The one with the widest distribution is the western lowland gorilla (*G. g. gorilla*), which is found in a small pocket of eastern Nigeria but has its main area from Cameroun through Rio Muni and Gabon into Congo (Brazzaville). This is the smallest of the subspecies, with standing males reaching a height of 1670 mm (66 in) and weighing in at up to 140 kg (310 lb), while females are about half this weight, a relationship which holds true for the other subspecies. *G. g. gorilla* may be distinguished visually from the other two subspecies by its nose, which has a continuous, heart-shaped ridge all round the nostrils, while in the others the ridge does not continue below the nostrils.

The eastern lowland gorilla (*G. g. graueri*) is found in a small area of eastern Zaire to the west of Lake Tanganyika, while the mountain gorilla (*G. g. beringei*) lives in a very small area of the Virunga volcanoes on the borders of Rwanda, Uganda and Zaire. The lowland form is on average slightly bigger than the mountain form but the latter has a longer, silkier coat, no doubt a necessity since it is found at heights up to 3000 m (10,000 ft). Adult males are

The two subspecies of orang-utan are restricted to the islands of Borneo and Sumatra respectively. This is a female Bornean orang (*Pongo pygmaeus pygmaeus*) with her offspring in Twycross Zoo.

often referred to as 'silverbacks' on account of the saddle of grey hairs which develop across the middle of the back. This tends to be less obvious in the paler-furred western subspecies than in the darker-furred eastern forms.

As most readers are probably aware, from having seen them on television, gorillas are almost completely terrestrial. Youngsters and females do venture into the trees at night, where they sleep in foliage nests; but the heavy males build nests on the ground because their large size precludes much arboreal activity. They are mainly vegetarian, though they will take small inverte-

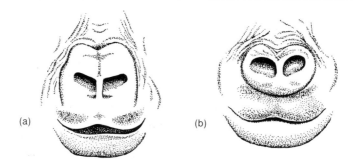

(a) (b)

Figure 10 The shape of the nose can be used to separate the lowland gorillas from the mountain gorilla. The nose on the left (a) is that of a mountain gorilla, while that on the right (b) is that of the lowland form.

The chimpanzee (*Pan troglodytes*) has a wide distribution across Africa from Senegal in the west to Tanzania and Uganda in the east. The way in which chimpanzees walk on their knuckles can clearly be seen in this picture.

brates when they find them, and being large animals they require a considerable daily food intake. The western lowland form eats a great deal of fruit, whereas the mountain gorilla consumes a wide range of plants and plant products. They are quadrupedal, walking on the flat of the feet and the knuckles of the hands. Social groups are generally based upon a single mature 'silverback' male with a number of immature males and mature females and youngsters. Newly mature males may be forced to leave the group and stay solitary until they can get together a group of females of their own.

The final pongids are the two species of chimpanzee, the common chimpanzee (*Pan troglodytes*) and the very rare and endangered bonobo or pygmy chimpanzee (*Pan paniscus*). *P. trogolodytes*, from now on simply the chimpanzee, has a wide distribution from Senegal in West Africa to western Tanzania and Uganda in the east, with its southern limit approximately across the centre of Zaire. The bonobo is found in scattered localities in one area of Zaire south of the rivers Zaire and Lualaba from which the chimpanzee is absent.

Chimpanzees are fairly large, with the males reaching about 50 kg (110 lb) and the females about 10 per cent lighter. Bonobos are considerably smaller, weighing a little over half what chimpanzees do; otherwise they are not easily distinguishable. Chimpanzees spend their time foraging both in the trees and on the ground, and although they are mainly fruit eaters, supplementing their diet with young leaves, nuts and seeds, they have been observed to catch and feed on monkeys and they also take termites and ants. When meat is on the menu it tends to be eaten more often by the males than by the females. Being smaller, bonobos are more agile than chimpanzees and spend more time in the trees. Anyone who has watched chimpanzees filmed in the wild will realize that, though basically quadrupedal, they also swing from their arms and on occasions walk bipedally. The social life of chimpanzees is more complex than that of the other apes, for though they forage in large groups these tend to be made up of individual subgroups consisting of males, or females with offspring of different ages, mixed groups and even solitary individuals. As they wander through the forests they may mix with other groups before separating once again, by which time there may have been some cross-migration between the two original groups. In some areas chimpanzee groups form territories and defend them against other groups. Because of the isolation of the region in which it lives, little is known about the ecology and behaviour of the bonobo.

Chapter 4
Reproduction and Parental Care

As with any other organism, the success of an individual primate species depends upon its ability to reproduce successfully and at least maintain its numbers between one generation and the next. In order to attain this end there has been a tendency in the primates to reduce the number of offspring produced by a single female, but then to ensure their development to maturity by means of a long period of parental care.

THE PHYSIOLOGY OF SEX IN PRIMATES

To all intents and purposes mammalian and thus primate reproductive cycles commence at the point in time when the female ovulates (oestrus) and is thus receptive to the male. The period at which oestrus occurs and the number of times it takes place annually vary considerably within the primates. In general, female prosimians ovulate either once or twice a year depending upon the genus concerned. The mouse lemurs, the slender loris and the lesser bush-baby ovulate twice a year, and as a consequence they have two distinct birth seasons. Most other prosimians, however, have a single period of oestrus per annum and a single birth season. The most extreme example of this is the ring-tailed lemur, whose mating season lasts for just two weeks; each individual female is in oestrus for just one day during this time. As a result all the young are born within a few days of each other. Alternatively prosimians show no seasonality, and in the indri, for example, females produce a single young every two or three years. In the mouse lemurs, the ruffed lemur, the fork-marked lemur and the ring-tailed lemur the female vagina becomes closed by an overgrowth of skin during the period of anoestrus. Tarsiers also have two periods of oestrus per annum.

The situation in the Anthropoidea is somewhat different, for in most of them breeding can occur at any time of the year and the females have both an ovulatory and a menstrual cycle, the latter being absent in the prosimians and also in most of the New World monkeys. One group of monkeys which do appear, at least in some species, to have a distinct breeding season are the tamarins of the genus *Saguinus*. The marmosets do not have a fixed breeding season but instead indulge in *post partum* mating. In *Callithrix*, for example, the females ovulate ten to twenty days after the birth of their offspring and in *Cebuella* after twenty-one days, at which time mating takes place. These species, therefore, are both pregnant and lactating simultaneously.

In the remaining anthropoids ovulation tends to be suppressed while the female is lactating, and she only begins to ovulate again once her offspring are

84

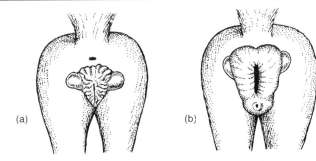

Figure 11 The perineal area of the females of a number of species of Old World monkeys and apes swells up when they are in oestrus. On the left (a) is the rear end of such a female during anoestrus, and on the right (b) the same female during her period of oestrus.

weaned. Much of our information on the length of primate reproductive cycles comes from research on captive animals, although where menstruation is visible as an indication of the timing of such cycles information has been gleaned from wild populations. As a result, little is known about the non-menstruating New World monkeys apart from *Ateles* spider monkeys, in which ovulation occurs every twenty-four to twenty-seven days, and *Lagothrix*, in which it takes place every twenty-one days.

In the Old World monkeys the ovulatory cycle lasts between twenty-eight and thirty-five days according to the species concerned. Humans, for example, share with the gibbons and some of the guenons a short cycle of only twenty-eight days, whereas in *Papio* and *Gorilla* the cycle spans up to thirty-five days. The record is held by the gelada, whose cycle lasts up to thirty-six days. The patas monkey shows an interesting departure from the other Old World monkeys in that it seems to have a definite birth season, which in its turn means a fixed mating season.

The receptiveness of a female when she is ovulating is demonstrated to the male in a number of ways. There may be an increase in pheromones secreted by the genitalia or by special glands; she may develop the areas of coloured skin mentioned in Chapter 1; or she may undergo behavioural changes so that she shows some degree of coquettishness. In some species only one of these ploys may be used, whereas in others combinations of two or all three of them are made use of.

The role of the coloured female genital area in primates such as baboons has always been assumed to be that of male attraction, but it is only recently that this has been proved beyond all reasonable doubt by experimentation. The subject chosen for this investigation was the chacma baboon, in which the area around the female anus, the perineum, swells a great deal a few days before ovulation. A number of females had their ovaries removed to neuter them and were then fitted with a prosthesis of red plastic to replace the normal swollen perineum. When such females were presented to sexually mature male chacma baboons a definite increase in the level of sexual activity was observed

The most common method of mating in primates is for the male to mount the female from behind gripping her ankles with his hind feet. This is a pair of chacma baboons (*Papio ursinus*) mating in savannah in South Africa.

in the males. What is more, there was a greater increase in the males' response as the size of the prosthesis was increased: there was a minimum size which elicited no response, while an exaggerated response was given to females with over-large prostheses. Since the neutered females would not present the males with any other form of attraction, it has to be concluded that it was the red prosthesis – and thus, in the wild, the swollen perineum – which altered the males' behaviour.

In many primates mating is typical of the quadrupeds, with the female presenting her rear end to the male so that he can mount her from behind. During mating the male may retain contact with the ground by his hind legs, or alternatively he may grasp the female's ankles in his hind feet which are thus raised off the ground. Gibbons and orang-utans are known to copulate face to face, even doing so whilst suspended by their arms from the branch of a tree, and face-to-face mating has been observed as an alternative to rear entry in capuchins and a number of other species.

There is not, as might have been expected, any positive correlation between size and gestation period in the primates. Admittedly the smallest of all primates, the mouse lemurs, have the shortest gestation periods – between sixty and ninety days according to species; but then the not much larger pygmy marmoset has a gestation period of 133–140 days and the similar-sized tarsier one of 180 days. At the other end of the scale gestation takes 264 days in the orang-utan and 258 days in the gorilla, although female gorillas are half as big again as female orangs. Only the spider monkeys come anywhere near these times, matching the gestation period of the chimpanzee at around 228 days. Within the guenons the savannah-dwelling *Cercopithecus aethiops* has a gestation period of 165 days, somewhat longer than that of 140 days for the other forest-dwelling members of the genus. It has been postulated that the longer gestation period in the ground-dwelling *C. aethiops* permits a much more rapid post-natal development of the offspring, so that it soon becomes capable of escaping from the dangers of a mainly terrestrial existence. This is not nearly so important for the other arboreal *Cercopithecus* species, where the young are in the relative safety of the trees.

The number of offspring produced at one time by female primates is low – one, two or on rare occasions three, according to species. Two or three is normal for the dwarf lemurs and the ruffed lemur, with the mouse lemurs producing normally two and occasionally three or just one offspring. Of the remaining prosimians the slow loris and the potto normally produce one but sometimes two offspring, while the dwarf bushbaby and the greater bushbaby always produce twins. The situation within the remainder of the primate groups is more simple. The tarsiers and all the catarrhines produce a single offspring, with twins as an occasional rarity in the great apes as they are in humans. Of the platyrrhines, the marmosets and tamarins always produce twins, with the exception of *Callimico*, which produces a single offspring, further evidence for its tenuous relationship to the other callitrichids. The Cebidae also produce a single offspring, although the capuchins and *Aotus* occasionally have twins. In the Old World monkeys and the apes a single offspring is the norm, though twins are born at about the same frequency as they are in humans.

Birth is a relatively rapid process in primates other than humans, and in all but the great apes, which are in little danger from attack by predators, always occurs during the night. From the small amount of data yet available it seems that the birth process in most primates takes about two hours although, as with humans, there tends to be some variation. Observations on chimpanzees, for example, reveal that birth can take as little as forty minutes but as long as eight hours, and whereas the process normally takes about two hours in gorillas, one female was seen to produce her offspring in just twenty-nine minutes. As in humans, extra-long birth times seem to result from some form of physiological problem. The record for size of babies in relation to that of the mother goes to the talapoins, for they weigh in at an enormous 20 per cent of the female's body weight. Put into human terms, this is equivalent to a 50 kg (112 lb) woman producing a 10 kg (22 lb) baby – some feat!

In the presence of the photographer this female long-tailed macaque (*Macaca fascicularis*), in rainforest in Java, puts a protecting arm around her offspring in much the same way as a human mother would if a stranger were present.

In many cases across a range of species some form of vocalization was noted in the females during the birth, and from the offspring following birth. Manual assistance by the female during the birth was noticed in a number of instances, and once the young was born it was quite often given a manual inspection. In nearly all cases the placenta was eaten by the mother once it was free of her body. This behaviour is of course typical of nearly all mammals, since it provides the mother with a nutritious meal and also hides any evidence from keen-nosed predators that there is a young and defenceless animal in the vicinity.

PARENTAL CARE

This is a very important factor in the primate success story, and as such it deserves to be considered at some length. Any female mammal, of course, cares for her young to some extent, since during the early stages of postnatal development she supplies them with much of their food in the form of milk. In many

The majority of young primates are at some time or another carried around by their mother and sometimes by their father. The female ring-tailed lemur (*Lemur catta*) at first carries her offspring hanging from beneath her belly, but as it grows older it climbs on to her back. This youngster is with its mother in its natural habitat in Madagascar.

of the primates, however, this period of lactation and care of the young has been extended to cover a much longer period than in most mammals. The advantages of parental care as typically seen in birds and mammals are that the young are ensured, at least under normal conditions, 'a ready supply of food; they are protected from potential enemies; and they have time to learn to look for food themselves and to discover the complexities of the society in which they live.

Since most primate births take place arboreally the newborn young has to be in a fairly advanced state of development and must be immediately capable of clinging on to its mother. The exception is found in certain prosimians, which produce their young in nests and care for them there for the first few days of their lives. Instead of clinging to their mother they are carried around in her mouth, in much the same way as a cat carries her kittens, until they are strong enough to be able to follow their mother around. In prosimians such as the potto, which does not make a nest, the young are left in a suitably safe spot while the mother forages. She then returns at intervals to feed her offspring.

From the age of about four weeks it is carried on its mother's back, but within another four weeks it is moving around independently. Once the young potto is on the move and weaned it learns from its mother what is good to eat. It does so by following her around and taking her food out of her mouth to taste and then eat. The female never seems to object to this, but on the other hand she never offers food to her offspring voluntarily.

There is some variation in the pattern of parental care in the lemurs. The young of mouse and dwarf lemurs, for example, are born in nests and remain there until they are able to move around under their own steam. Observations on a wild population of the ruffed lemur in Madagascar revealed that females build their nests in large trees at a height of between 10 and 20 m (30–60 ft) two to three weeks before the young are due to be born. These nests are always well concealed amongst thick tangles of foliage and are constructed from slender, leafy branches. Two weeks after birth the offspring are moved out of the nest and their mothers carry them in their mouths to a parking place in the trees. This is usually on flat areas amongst tangles of lianas and other plants, where they feed; twins are usually parked together, although offspring from different mothers are always placed in separate locations. The infants are moved at intervals, from one to five times per day during their first month of life and between three and seven times during the second month. The young-sters are left on their own for hours at a time while the mothers go about their various activities, though in the first month of their life the separation between the two is no more than one metre for more than half of the day. As the infants grow, however, the separation distance increases and the percentage of time spent by the mother in close proximity gradually decreases. As the mobility and independence of the young increase, so the mothers spend less time in caring for them and more in other social activities within the group.

By the time the young are two to three months of age they are able to follow the adults in the group for considerable distances through the forest canopy, though of course rather slowly, and on occasions the mothers will sit and wait for their offspring to follow. As long as they are small enough to fit in her mouth the female will occasionally help them to cross gaps, but once they are too big for this they have to make their own way.

The young of *Lepilemur* are usually born in a nest in a tree-hole and left there by their mother while she forages for food until they become mobile. *Hapalemur* has a similar pattern of care, but in this case the young are occasionally carried by the mother. In the remaining lemur genera, the young are carried first on the mother's belly fur and then, when they are old enough to do so, on her back. The avahi male shares the carrying of the offspring, and this is also true of the red-bellied lemur. Males of this species carry the young at around three months old, when they are still suckling but can move around independently. Seemingly, when they get tired they hitch a ride on their father instead of on their mother.

Tarsier young are born at an advanced stage of development, and for the first three weeks of their life are carried around by their mother. As with some of the prosimians, the baby is held in the mother's mouth and parked in some suitable position while she hunts for food. By the time it is four weeks old, however, the young tarsier can already hunt for its own food.

The New World marmosets and tamarins of the family Callitrichidae typically live in extended family groups of a single reproducing male and female and their offspring of different ages. Usually twins but occasionally triplets are born twice a year, the newborn young having to cling on to their mother's fur and make their way to a nipple before she begins to take any notice of them. Initially the young are carried by the mother, but within a short time the male takes over this duty. As they get older, and with the birth of following sets of twins, so the older offspring play their part in caring for and carrying their younger siblings.

Whereas female monkeys are always prepared to part with milk to feed their offspring, the adults in general are usually reluctant to part with food items. This is not, however, the case with the marmosets and tamarins, for as the young begin to take solid food this may be provided for them in premasticated form by their father. Alternatively they may share an item of food held in their mother's mouth. It is now known from observations on a number of different callitrichid species that the youngsters will actually beg food items from adults. Such behaviour has been well studied in a group of free-ranging buffy-headed marmosets in Brazil. It was discovered that if one member of the group found a particuarly large and desirable food item, then attempts would be made by other monkeys in the group to obtain it. Except in the case of the reproductive female or youngsters of up to eight to ten months, the possessor of the food item would not tolerate these attempts to hijack its meal. The youngsters would beg any kind of food item, and it was observed that on occasions their mother would accompany them and support their efforts to solicit food items from the other members of the group. Most interesting of all in these studies was the discovery that, if the oldest members of the group found an especially large and succulent insect, they would actually try to attract the attention of the youngsters to it. This possessor-initiated food transfer behaviour has also been observed in two species of tamarin.

In order to attract the attention of the infants, the possessor of the food holds the item out in front of its body and then gives a call very similar to that of the youngsters' own begging call. On hearing the call, the young marmoset then replies with a high-pitched chattering whistle which it continues to produce until it reaches the caller. Once there the infant is allowed to take the food item from its possessor. On occasions when a youngster was near enough to be able to see the food item clearly in the first place, the call was dispensed with.

Another group which has been studied extensively in the wild is the yellow-handed titi monkey. Males and females form lifelong pair-bonds, and the groups comprise the couple and their offspring of various ages. The titis produce only one young per year, and they remain within the family group for two or three years before departing to form partnerships of their own with a monkey of the opposite sex. Parental care in these monkeys is of interest in that it is carried out almost exclusively by the male. He takes over transport of his offspring from its second day of life, and only passes it back to its mother when

The family group is an integral part of the lifestyle of marmosets. Here a buff-headed marmoset (*Callithrix flaviceps*), in Brazil's Atlantic coast rainforest, is carrying a number of youngsters on its back.

it needs to suckle. The male carries the youngster on his back, but it has been observed that during heavy rain he will shelter it beneath his body. This paternal attention was found to increase as the youngster grew. Fathers were seen to groom their offspring and stand guard as they played or foraged for insects on the ground. The male titi also shared his food with his offspring, though the female never did so.

An interesting piece of information concerning parental care has been obtained from female vervet monkeys. Over a period of time a group of females who had failed in their previous pregnancy were compared with successful mothers or mothers with their first babies. It was found that with their next baby the original failed mothers encouraged more ventral contact with it, restricted its movements more, gave it more care and attention than the other two groups of mothers, and played a greater part in keeping the youngster near to them. What is more, in the females who had failed in their preceding pregnancy there was an increased interval before they became pregnant again, as if some internal clock had been adjusted to make sure that they made a good job of raising their new infant after losing the previous one.

Parental care lasts considerably longer in the apes than it does in any other primate group. Gibbon young, for example, do not leave their parents until they are seven or eight years old; as a result, since babies are born every two or three years the gibbon family is not unlike a human family. The young of the other apes spend a similar length of time with their parents before becoming independent. During this period they learn how to forage, how to recognize and escape from their predators, and the rules of the social groups in which they live.

Alloparental care

So far we have discussed care of infants by their parents or by their older brothers and sisters. Within the primates, however, there are many examples of youngsters being cared for by adults, either females or males, who are not immediate family. It is to this phenomenon that the name of 'alloparental care' is attached.

Alloparental behaviour has been observed in both prosimians and simians, and in many is an integral part of their social groupings. In the group of ruffed lemurs whose parental care was discussed previously, alloparental care by an adult male and by two females without infants was also observed. These alloparents displayed such activities as sitting in contact with other lemurs' babies, grooming them, playing with them and following them closely as they travelled around. What is most interesting about this specific group is that the youngsters actually spent more of their waking hours in contact with the adult male than they did with their mother or the other females – perhaps because he could well have been their father. The workers who observed this behaviour in the wild suggest that its importance lies in the fact that it frees the young lemurs' mother to forage and thus to increase the amount of food available to them. In other lemurs it is not normal for a male to act as a babysitter, but the phenomenon has been observed in a large group of Verreaux's sifakas. In this instance a lone adult male was observed to carry his females' infants around with him. This presumably reduced the burden of care placed on his females, with all its concomitant advantages. Why such a behaviour pattern has not evolved as the norm in these creatures would seem something of a mystery; perhaps one day all male sifakas will indulge in such activities, rather than sitting around all day doing nothing in particular.

A somewhat unusual example of allomothering was observed when a group of three unrelated female ring-tailed lemurs was released into a forest enclosure. One of them became pregnant as a result of a male accidentally gaining access to them from an adjacent enclosure. She eventually gave birth to twins, and two weeks later one of the other females began to carry the twins around, one at a time. When the females were examined one month after the birth it was found that not only was the twins' mother producing milk, but so also was the unrelated female who had been seen to carry them around. The third female at no time carried the infants, nor did she lactate, but she did help to defend them from the attentions of wandering males whenever they approached the twins.

A special kind of alloparental care, which lies outside the normal behaviour patterns related to the social life of primate groups, is that of adoption, where an infant or infants change from one primary caregiver, their mother, to another individual outside the family, not necessarily an adult female. True adoption can only happen to an immature individual, and does not include instances where a mother dies and her place is then taken by other members of her immediate family, as would happen in marmosets for example.

If an infant loses its mother, whether permanently as a result of her death, or temporarily due to illness, it requires a source of milk if it is to survive. Adoptive suckling of such infants has been recorded for many primates in captivity and in the wild for howler monkeys, several langurs and macaques, and the mountain gorilla. For example, in wild chimpanzees it was observed that a young male was separated from his sick mother for six days and during this time he was looked after by two non-lactating females who provided for his every need, except of course milk. He eventually went back to his mother. In captivity a lactating female, with an infant of her own, for several hours looked after the four-week-old baby of another female who was ill. Where the adopter is, however, a non-lactating individual then the chances of an unweaned infant surviving are very small. For weaned infants with a long development time to maturity, adoption is still quite important since an immature primate requires such things as warmth, transport and comfort as well as food. This kind of adoption also provides a degree of protection for the infant from an experienced adult. Putting aside instances where male adoption of young females is part of the species social structure, even male primates will on occasions adopt immature individuals, though reports on how successful they have been are somewhat contradictory: some observers report them as being competent, while others claim the opposite to be true.

Cases of adoption of infants abandoned by apparently healthy mothers have been recorded. Sometimes young are sneaked away from their mothers by a female who little by little gives more and more care to the infant until she eventually becomes its primary caregiver, with little or no opposition from its true mother. Adoption must, however, be seen as separate from kidnapping, in which an infant is taken from its mother without her consent and she protests accordingly. Kidnapped infants are normally less than four weeks old, and the deed is usually carried out by another female who is typically dominant over the mother or might come from a different social group.

SEXUAL BEHAVIOUR PATTERNS IN PRIMATES

In all but a few species of primates, sexual behaviour is inextricably tied up with social groupings. In those non-social species where males and females mate for life the situation is quite straightforward, for the male can copulate with his mate whenever she is in a receptive state. Prior to actual mating there may be a period of courtship, such as that which has been observed in tarsiers. Here females in oestrus take an active part in initiating courtship by opening their legs and displaying their swollen, red vulva, which stands out starkly against the white inner thigh fur. This act is performed a short distance from the male, who responds with a chirruping call before approaching and sniffing the female's genitals and twigs which she has previously urine-marked. The male then urine-marks the spot himself, to which the female responds by jumping away. This pattern of behaviour is repeated at intervals of ten to fifteen minutes until between sixty and ninety minutes after the beginning of courtship mating takes place.

As the young of these non-social species mature, they leave family groups and seek lifelong mates of their own. In multi-male groups or those groups based on harems, however, the situation can be much more complicated. Ring-tailed lemurs and sifakas both live in multi-male groups which are dominated by the females in it. They have slightly different strategies, for in ringtails the group males mate with the group females but at intervals they change groups, whereas in the sifakas the males tend to seek out females from other groups during the mating season. Both ploys, of course, increase the chances of outbreeding with all its advantages.

Solicitation of males by females appears to be the norm in many anthropoids, though in a number of species this may take place outside periods when they are actually in oestrus. An example of such behaviour is to be found in forest-dwelling hanuman langurs. Solicitation, during which the female points her rump towards a male, at the same time rapidly shaking her head sideways, takes place throughout the year. However, mating, when presumably the females are in oestrus, only takes place during the monsoon season and hot weather, mainly from April to August but also during October. At no time has mating been observed to have taken place without a preliminary solicitation of the male by the female.

Whereas in most monkeys the females mate either with their lifelong mate or with the dominant male of the group, in barbary macaques the females are highly 'promiscuous' and mate with a number of males in rapid succession. It is the females who solicit the males, and as their time in oestrus increases so they mate with more and more males rather than repeatedly with the same one. This is in stark contrast to the dominance hierarchy systems found within the other macaques, where the dominant male gets most of the mating opportunities. This apparently unusual behaviour of barbary macaques seems to relate to the fact that all the females ovulate synchronously, a phenomenon apparently resulting from ecological constraints which dictate that it is most advantageous for babies to be born in the spring over a relatively short birth season. With all the females in oestrus at the same time the males are unable to

Male baboons fight for the right to mate with receptive females. Here a male chacma baboon (*Papio ursinus*) is barking at a rival in mopane-tree veld in South Africa.

defend their own individual partners, and as a result they mate with any female at every opportunity that arises. One great advantage of the system appears to be that it reduces male–male competition, so that the energy wasted by competing males can be better employed elsewhere, such as the care of young born the following spring.

In multi-male capuchin societies, who mates with whom is dictated by genetic relationship, social status and also by the fact that in some instances particular individuals show a preference for each other. Courtship is quite intricate and it involves vocalization, posturing, gesturing, facial expressions and sometimes play – if the male does not show sufficient interest in the female, she may slap or pull at him.

Because they are highly visible it is much easier to study terrestrial primates in the wild than it is to study arboreal species, and as a result we know a great deal about the sex life of the baboons and their kin the geladas. Hamadryas' sexual strategy, for example, is based upon the harem, which remains in the sole ownership of a male for as long as he is fit enough to maintain it. Young males begin forming their own harems by kidnapping and adopting weaned females as young as two years old. The male then looks after the female, at the same time teaching her to groom him and keeping her away from other males. At first these pairs may consort with pairs of similar age, but gradually the male may adopt other young females and, as the size of his harem increases and his first female reaches sexual maturity, so these relationships decline. Once she is old enough to mate, the female is kept away from other males by physical punishment such as a sharp nip on the back of the neck; she soon learns to return to her 'lord' in order to avoid further punishment. In these circumstances the harem master is always on hand to mate with his female whenever their coloured posteriors indicate that they are in oestrus. Despite the fact that they are herded together all the time, there is little bonding between the harem's females.

This is not the case in the harems of the gelada, for in this species the females form tightly bonded relationships so that they do not have to be constantly herded together by the male. On the other hand he has to fight to form his harem in the first place, either by defeating the resident male of a ready-formed harem or by milking off young females from other harems and gradually building up his own. The remaining baboon species within the genus *Papio* do not have harem-based societies but live in multi-male groups where the dominant males fight for the right to mate with the females when they are sexually receptive.

Within the apes, the gibbons live in family groups where the male and female are mated for life, while orangs tend to live solitary lives though they are occasionally seen in rather loosely connected groups. Under normal circumstances males will form a temporary relationship with a female who is approaching oestrus and will then mate with her at the appropriate time, whereupon he will resume his solitary existence. Gorillas live in groups dominated by a silverback male who mates with receptive females. In common with chimpanzees, who form multi-male groups, outbreeding is maintained by the movement of females away from the groups in which they were born and into another group.

Chapter 5
Social Behaviour

Living in some kind of social grouping is not unusual among mammals and is certainly not special to primates. Wolves, African hunting dogs and lions all exhibit complex intra-group relationships whose sophistication may in some aspects rival or even excel that seen in primates. Even vampire bats demonstrate a form of social behaviour and of a type rare among primates – the voluntary sharing of food. A vampire returning replete with blood may regurgitate some of this liquid largesse into the begging mouths of its less fortunate (and probably unrelated) fellows back at the roost. Such apparently selfless generosity to non-kin is rare among any animals and very unusual in primates (although with the vampire bat it may not be as selfless as it initially appears, as today's blood-giver may be tomorrow's receiver). Even such an unlikely animal as the naked mole-rat exhibits strangely insect-like habits. It has a complex ant-like caste system in which the sole breeding 'queen rat' suppresses breeding in her followers, who devote their lives to helping in the rearing of a long line of her offspring.

Primate social behaviour resembles neither that of the vampire nor of the bizarre mole-rat. In fact in many cases the present lack of field studies means that we are uncertain just how the social systems of many species actually function; but increased human presence in the field is rapidly yielding much more information on this and other topics.

Social behaviour is based on the 'home range', the area which is regularly used by the animal or group of animals for all their regular activities. The home range may not be exclusive, but may be shared with other individuals or groups of the same and other species. Certain parts of the home range may be used more than others – some primates only regularly use the 10 per cent or so containing the richest food resources and favourite sleeping sites. The borders of a home range are not defended, but those of a 'territory' most certainly are. Territorial primates use many methods for guarding against trespass across their boundaries to ensure that they have exclusive use of their reserved living space.

At its simplest, primate social behaviour may appear at first sight to be non-social, for many of the nocturnal prosimians appear to lead noticeably solitary lives. However, just because a weasel lemur or bushbaby is almost always encountered feeding alone at night does not mean that it lacks regular communication with others of its kind. There are two main methods of ensuring that the neighbours know who is around, while at the same time possibly disclosing intimate details about sex and breeding condition: by shouting loudly about it, and by leaving smelly calling cards. The latter method is

Some primates appear at first to lead solitary lives. The weasel lemur (*Lepilemur mustelinus*) spends the day in or beside a tree-hole, foraging alone at night within its small territory but keeping in regular contact with its close neighbours via frequent loud calling. Photographed in the reserve at Bezah Mahafaly in the dry south of Madagascar.

probably employed by all nocturnal prosimians and has the advantage of staying-power. It is also less risky, being a private communications channel decipherable only by the noses of other members of the same species and therefore less likely to attract the unwelcome attentions of predators than when sitting around in an exposed spot and calling loudly. If applied in a particular way a scent may be remarkably enduring, as in the *Cheirogaleus* dwarf lemurs which use their extruded anus as an applicator to smear a line of faeces along a branch, leaving an odorous 'squatter's rights' proclamation capable of sticking around for many weeks or even months. Bushbabies also leave scent messages, but use a multiple applicator in the form of the hands and feet. These are meticulously daubed with urine before the animal moves off, leaving at every step a gradually declining trail of smelly footprints which no doubt reveal a mine of information to the recipient.

Like most (perhaps all) nocturnal prosimians, bushbabies also use sound – up to eight different calls have been noted in the lesser bushbaby – to communicate with their neighbours; the calling of these diminutive animals can be one of the most characteristic sounds of the African bush after dark. Bushbaby society generally consists of several overlapping female home ranges which are more or less completely covered by the much larger home range of a single male. He makes it his frequent business to visit each female's range to check out her breeding condition, to be certain of being in the right place at the right time to mate with her if she is in oestrus. Although he may not be the only male in the vicinity, he is the largest and toughest, so probably secures sole access to the females. Rather more lightweight males manage to squeeze in a living between the ranges of the females and the central 'boss' male, while other males who have just reached adulthood lead the itinerant life of the drifter, passing at will through the ranges of both females and central 'boss' males. The presence of even smaller immature males may also be conceded within the range of the central male, presumably because they do not currently present any kind of sexual threat. So although bushbabies always seem to be alone when located in the forest at night, they do lead busy lives which include regular social contact with one another.

Horsefield's tarsier probably has similar arrangements in the forests of Borneo. Adult males seem to have larger home ranges than females, from around 8.75 to 11.25 ha (21.6 to 27.8 acres) for males and 4.5–9.5 ha (11–23.5 acres) for females. Unlike the galagines, whose females often sleep together during the day in a communal nest, both sexes of Horsefield's tarsier sleep alone. In fact all the evidence suggests that this species is perhaps the most solitary of all primates, regularly communicating with others of its kind only through calling and scent marking. However, there is a broad repertoire of calls, some rather insect-like, which serve to keep close neighbours aware of each other's presence. This may escalate to the point where at certain times veritable concerts may be struck up by several animals (up to five) perched in close vicinity. These concerts may be remarkably enduring, and in one case lasted for four hours. This concert style of calling seems to be quite different from the alternate duets found in the spectral tarsier and in unrelated animals such as the weasel lemur. The precise function of these tarsier concerts is unknown, but they may be connected with territorial boundary disputes. The exact nature of male–female relationships in Horsefield's tarsier is currently unknown; but it seems probable that, unlike in the galagines, two or more males may regularly visit the range of a single female from their own ranges, which only overlap with hers to a marginal extent. Scent marking may not be important and vocalizations are probably the main form of communication.

The social life of that most aberrant of primates, the aye-aye, remains something of an enigma, although it does seem likely that aye-ayes are loners and any female who dares to encroach on a male's favourite feeding tree will probably be chased away unless she is on heat. Aye-ayes spend the night sleeping way up in the canopy in a nest of twigs, and the first task after emerging at dusk is to devote several minutes to giving the fur a thorough grooming, a task which is usually performed while the animal is hanging suspended upside-down by its feet. Aye-ayes sleep and forage alone for most of the year, although at certain seasons several animals may suddenly start sharing a nest in an abrupt outbreak of togetherness. It is quite common for several different aye-ayes to feed in the same trees on the same night, so perhaps home ranges show considerable overlap.

In several nocturnal prosimians sleeping huddles are common among the more sociable females, but the strictly solo life is the norm for the males. Such solitary habits are not found again among the diurnal lemurs and monkeys until we come to the virtual top of the scale with the orang-utan, one of the most advanced of primates yet with social habits not unlike many of the more 'primitive' examples of the primates.

Orang-utan populations fall into three distinct categories: adult males leading very solitary lives; adult females, mostly with one or two dependent offspring tagging along and sometimes associating with other similar groups for a day or two; and sexually immature animals of both sexes, also mostly leading a lonesome existence. In brief accidental encounters between adult males there is a general disinclination to seek further mutual contact and a strong tendency to shy away from one another; such fortuitous meetings are very rare, probably because the males' frequent renderings of their 'loud call' may help to prevent their coming too close to one another. However, if a female is present in consort with a male then the situation may take a very different and much nastier turn, sparking off a lengthy and deadly serious tussle disputing access to her sexual favours before one or other of the males gives up the struggle and makes off.

In fact most orang-utan social behaviour, if the term is really justified at all, takes place very much in a sexual context. A female in oestrus will probably quickly attract the attention of an adult or subadult male who will establish a consortship with her, perhaps leading to repeated bouts of copulation over several days. In common with most human youths eager for early and satisfying sexual adventures, subadult male orang-utans are not over-fussy about their partners and will enthusiastically copulate with any female who shows willing and sometimes with those who don't – hot-blooded young male orangs tend to be somewhat forceful and reluctant to take 'no' for an answer, so an uncooperative partner may simply end up getting raped. This may be accompanied by anything from anguished squeals of protest to full-blooded battles of resistance.

Adult males, on the other hand, are more discerning in their appreciation of female values and less liberal with their attentions. So even some eager young seductress flagrantly exposing herself open-legged before his seasoned but disinterested gaze may fail to stir him. He is simply past the stage of temporary liaisons with all-comers which typified his youth, and only interested in investing the necessary time and effort into consorting with well-matured and experienced females capable of making a good job of rearing his offspring. When such an occasion does arise he will stir his great bulk to fierce defence of his investment. Thus the apparent lack of libido in these big old males and the seemingly early demise of their sexual powers may, it seems, merely be the mature development of a finely honed ability to judge feminine qualities.

Such carnal sagacity also extends to the actual act of mating, for whereas subadult males often 'try it on' with a female who is patently and often vociferously unwilling and end up taking her by force, adult males usually only consummate a relationship with fully adult females who openly solicit their attentions. This may be important, as in many mammals rape is thought to result in a far lower rate of pregnancies than is likely in copulations which have been actively encouraged by the female.

One step up from the 'solitary' lifestyle of frequent vocal or olfactory contact but little actual physical interaction is the 'family group' found, but only sparingly, in a wide variety of unrelated primates. A family group usually consists of a reproductively active male and female accompanied by one or more of their offspring. Among the small nocturnal species this lifestyle is found in the spectral tarsier from Sulawesi, a group of which will inhabit a remarkably small home range of only 1 ha (2.5 acres) or so. Territorial considerations play a significant part in their lives. Scent marking by the males is noisily backed up by calling and at least fifteen different calls have been distinguished, many of which are used in shouting matches with opposing neighbours. It is in this species that the 'duet for male and female voices' forms part of the territory-delineating song repertoire.

Long duets deafeningly performed by both sexes are also the main form of territorial proclamation in the gibbons. These long-limbed apes live in monogamous family groups of a male and female along with several of their offspring. The juveniles span several years of reproductive effort, constituting a type of social system which probably stems from the gibbons' likely habit of pairing for life. Their song is remarkably sophisticated and elaborate, as well as being extremely loud. The 'great call' of the large siamang overcomes the smothering effect of the billions of leaves in the forest canopy so successfully that it can be heard at a distance of 1 km (1100 yds) or more. During the elaborate duets generated by Mueller's gibbon (*Hylobates muelleri*) the males and females produce their own individual songs which interact in an intricate interlacing of sounds. These show a distinct sequential organization in which the female always takes the lead. Solo singing is restricted to the males, who give vent to lengthy and involved bouts of song which lacks the sequential arrangement heard in the duets. Each species of gibbon has its own particular song, which is recognized by opposing groups and helps stave off boundary disputes.

However, singing alone may not always adequately serve this purpose and all too often a border encounter deteriorates into blatant displays of aggression

The largest of the lemurs, the indri (*Indri indri*), forms small monogamous family units which employ a deafeningly loud chorus to maintain territorial integrity. Indris are not very sociable animals and physical contact between group members is infrequent: mutual grooming does not have the importance with which it is regarded in many primate societies. Photographed in the reserve at Perinet in Madagascar.

or even physical combat. When this happens it is always the males who go in for the rough stuff, while the females sit and watch or even indulge in a spot of mutual grooming. It seems that, as in human arguments, the initial 'war of words' between conflicting gibbon groups may not always suffice to avoid the outbreak of battle. Gibbons need large territories to provide them with the wherewithal to survive through the leaner seasons. These are around 54 ha (133 acres) in the lar gibbon and around 48 ha (119 acres) in the siamang which, though larger, tends to cover less ground during each day. With the omnipresent chance that pushy neighbours may try their luck at trespass and the possible annexation of a slice of territory, it is not surprising that boundary defence occupies a remarkable percentage of a gibbon family's daily routine. It involves singing on most mornings (and sometimes in the afternoons as well) plus fairly regular border skirmishing, although this is very rare in the siamang whose earth-shattering outbursts of afternoon singing generally seem to be enough to keep any rivals at a respectful distance.

The song of a gibbon family echoing through the dripping groves of stately tree ferns and moss-draped trees in some Malaysian rainforest is undoubtedly both impressive and memorable, but for that special quality of spine-tingling eeriness calculated to raise the hairs on the back of the neck it cannot compare with that master among primate songsters, the indri. When an indri family's wailing chorus erupts suddenly and without warning from just a few metres above one's head in some damp Madagascan forest the effect is heart-stopping.

The indri is the largest of the extant lemurs, looking surprisingly like a big black and grey teddy bear. Indris have a remarkably laid-back lifestyle, rising late, retiring early and not moving very far or very fast in between. They live in monogamous family groups of two to five individuals, usually comprising a male and female along with their offspring. These may occasionally be joined by one or more subadults from another group, their presence apparently being tolerated mainly because their eventual intention is to pair off with the resident juveniles when they become mature. The family territory usually occupies 15–30 ha (37–73 acres) of upland rainforest and has a perimeter band around 30–50 m (40–60 yds) wide which is shared reasonably amicably with neighbouring groups. Border disputes are rare, probably because each indri family's regular song sessions serve adequate notice not only of their continued ownership of that patch of forest but also of the sex ratio within the group, its size and possibly even the sexual condition of its members. The latter possibility has been inferred from the considerable increase in singing noted during the breeding season in midsummer.

For the indri song represents a particularly useful method of social communication. With its rather low-quality leafy diet and fairly cool habitat the indri would find regular aggressive defence of the territorial boundaries a most tiresome and energy-wasting affair. Song is less energy-demanding, and yet by penetrating a long way it has the advantage of conveying information to several groups at the same time without the need even to move a limb – let alone start a fight. Indeed, the indri's song is truly sensational in its power, carrying for up to 2 km (1¼ miles) under favourable conditions, during which it may reach the ears of as many as twenty different groups round about. Song also virtually eliminates the need for regular demarcation by scent marking, a habit much less common in the indri than in almost any other lemur.

It is the male who first strikes up the indris' weird chorus, this being just about the only occasion when he takes the lead in anything. In all other respects he is thoroughly subservient to the female, to whom he meekly and obediently gives way in everything – he will even give up a favourite feeding spot to her, usually without the slightest sign of hesitation or resentment. Such placid reactions are typical of the indri, and female-led groupings are not unique to them; in fact they are so common in the lemurs as to be more or less the rule. This is an important aspect of lemur social organization which sets them apart from the mainly male-dominated monkey and baboon societies.

Watching indris for hour after hour in a Madagascan forest can be a rather patience-stretching and motive-questioning experience. If the weather is wet,

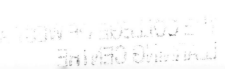

droves of ravenous leeches eagerly loop their way across the damp leafy carpet to latch on to the tasty human. Unfortunately the human is rendered easy and very static meat through being forced to sit there for long periods just waiting for the typically somnolent indris actually to *do* something in the trees above. However, even baby indris are imbued with the playfulness natural to most infants, so if a youngster is present in the group it may at least try to liven things up by trying to instigate a bout of play with its parents, scrambling around over their unresponsive bodies as it attempts to elicit some reaction. Alas, all too often such adolescent high spirits are doomed to failure in the face of the uncompromisingly unresponsive constitution of the average adult indri. Even mutual grooming, the delight of so many primates, is rare among indris, in whom any kind of physical contact seems in general to be shunned.

In some parts of its range the indri may live alongside another species of lemur boasting considerable stentorian powers, the beautiful ruffed lemur. Smaller and lighter than the indri, and considerably more agile, ruffed lemurs live in pair-bonded groups of a male and female with their offspring. The 'bark-bray' call, somewhat reminiscent of a cross between a dog and a donkey, suddenly erupts from the foxy snouts of these impressive animals with a startling unexpectedness which is particularly acute when it happens at night, in marked contrast to the exclusively day-singing indri. For an animal no bigger than a large domestic cat, the volume of sound produced is really staggering. Every other sound in the forest is drowned out as a calling round between several neighbouring groups is set in train. Such rounds of song initiated by one group and taken up by others, often far into the distance, are also typical of the indri, certain gibbons, titis, howlers and the guereza.

The ruffed lemur has a wide vocabulary of calls for use in a variety of situations, from the bellowing 'roar-shriek' which dominates the forest in communication between groups to a rapid chatter signifying submissive status and a soft mew normally exchanged between a mother and her infant; in all there are thirteen different sounds. As in the similarly loud-mouthed indri, ruffed lemurs seldom if ever use scent to demarcate territory. The small family groups of the red-bellied lemur from the same rainforests are strictly territorial, fiercely defending the family boundaries in rowdy fights with neighbouring groups. Both sexes delineate the perimeter of their territory by scent marking with the ano-genital region, although males also deploy a secondary marker from a gland situated on top of the head.

A rather sophisticated 'mutual-aid' style variation on the monogamous family unit is found in the tamarins and marmosets: all members of the family, from father to mature offspring, take part in the care of the newborn young. As this is such a fascinating aspect of primate reproductive behaviour it is described in greater detail in Chapter 4. Another interesting slant to the lives of the callitrichids is their heavy reliance on scent marking to define territory, a custom which is rare or absent in most Old and New World monkeys but common among the Madagascan lemurs. Saddle-back tamarins apply scent marks to lianas and slim branches using three different methods: ano-genital, by sitting upright and rubbing their backsides against the branch with a rotating action, possibly at the same time reinforcing the message with a spattering of urine; suprapubic marking, in which the animal lies prone and pulls itself forwards with its arms, massaging the suprapubic region against the bark to smear it with scent; and, much more rarely, sternal marking, in which the chest is rocked to and fro against the bark. As in lemurs, most of this activity is concentrated around the periphery of the home range. Scent marking may be applied more intensively on a number of favourite spots, such as a certain liana which may attract a thorough going over from the whole group in concert.

Similar multi-faceted scent posts are also established by the ring-tailed lemur, whose social arrangements are based on the so-called 'multi-male' system; although in this lemur, with its female-led society, the system works in a slightly different way from the generally male-dominated groupings of monkeys. Ring-tailed lemur females and their recent offspring form the stable nucleus of their group, partly because the females are the bosses and partly because they are a permanent feature of the group until death parts them from it. The males by contrast are more footloose, leaving the group of their birth and trying their luck with another one, before perhaps quitting that as well and joining another or maybe even a succession of groups. In the process several of these rootless males may temporarily meet up to form an all-male group, but this seldom lasts long before the members leave to join a group which offers the possibility of an all too brief burst of sexual activity because it contains some females.

In ring-tails there is a definite dominance hierarchy within the group. Males give way to females on all issues as an accepted matter of course, but the females have their own dominance hierarchy among themselves in such important considerations as access to the best feeding sites. However, the practical functions of female status are restricted to such matters and rank is not allowed to spoil the real pleasures of life. For a ring-tailed lemur the greatest pleasure of all is an extended session of mutual grooming in which any female, no matter how junior in rank, can freely approach and groom the top female without fear of retaliation; the dominant female is just as generous with her own attentions, so that a low-ranking member of the group has an equal chance of being at the receiving end of the number one's toothcomb. This activity seems to rate higher in importance to ring-tails than to any other lemur and grooming pods of several animals are often formed, all with their faces buried in each other's fur as the dental comb is put to work. The young juveniles are also the object of considerable attention, and the chance to give a reluctant baby's fur a thorough licking over will be instantly snapped up by any member of the group; human babies, it seems, are not unique in being attention-grabbers. As they grow up, the juveniles themselves soon start to court one another's company, forming play groups featuring rumbustious bouts of rough and tumble on the ground or breathless acrobatics dangling from a loop of lianas in a frenetic game of arboreal tag.

Play is probably important in most primates, for it is at this early stage that social relationships may be established and alliances formed which may be

Long sessions of mutual grooming form an integral part of the social system of many primate societies, and it is typical to see monkeys picking meticulously through one another's fur. The function of such activity seems to have more to do with social bonds than with purely hygienic considerations, and in many monkey societies the dominance hierarchy dictates who can groom and be groomed by whom. These are long-tailed macaques (*Macaca fascicularis*) in a Javan forest.

remarkably durable. Sometimes the juveniles of one species may try to get a game going with those of a different species, although not always with success. For example, in The Gambia in late March juvenile grivet monkeys (*Cercopithecus sabaeus*) are quite well grown and no longer being carried by their mothers, but the red colobus babies are still quite young and spend most of their time clasped against their mothers' undersides. During group resting periods the red colobus babies, being still very young and more playful, often rush around trying to entice the grivet youngsters into joining in some fun and games. The grivets tend to give this the cold shoulder, but the red colobus babies go on trying just the same.

Play is not, however, restricted to juveniles. Adult sifakas often indulge in games of 'footsie', kicking playfully against one another or briefly engaging in a wrestling match while hanging precariously beneath some lofty branch in a gravity-defying display. Milne-Edwards diademed sifakas even have specific 'playgrounds' – special bare patches on the forest floor where there is plenty of room for a spell of energetic frolics without getting caught up in the vegetation.

During these sifaka playtimes the participants, adults as well as young, turn somersaults, vault over one another, engage in scrappy grappling matches, roll over and over and get covered in dirt and bits of dead grass and leaves, rush back and forth in a crazy game of tag and generally have a thoroughly good if rather exhausting time. Things quieten down every now and then as the rather breathless sifakas stop for a brief rest, sitting splay-legged on the ground before one of the more impetuous of their number rushes up to a playmate and sets off another round of hectic gambolling. These playgrounds are used so regularly that the earth may become smoothed by the frequent passage of tumbling bodies, and experienced sifaka-watchers know exactly where their group is heading when they suddenly take off rapidly in the direction of the playground.

Returning to the ring-tailed lemur's social arrangements, the dominance hierarchy within the males has a greater effect on the individual's day-to-day behaviour than happens in the females. High-ranking males swagger along with tail curved up in an S above their backs, but subordinates tend to dawdle along at the rear of the group with head and tail held low. However, even male ring-tails are generally quite nice to one another and do not take considerations of rank to extremes, so there is none of the cringing 'life of hell' suffered by very low-ranking members of certain species of monkeys such as the toque macaque. There is an even greater divergence regarding access to the females, for in monkey society it is almost always the top dog or alpha male who enjoys exclusive conjugal rights, at least if he can ensure it, whereas among ring-tail males the whole rank system collapses entirely when the females come into their incredibly brief period of heat. Just who gets to mate is decided not by previous rank but by who can come out on top in the savage fights which disrupt the harmony of ring-tail society at this time.

Ring-tail territories share a narrow area of overlap which can be freely entered by members of opposing groups. This common-user zone is the centre of attention for each group's scent-marking activities. In ring-tails the nose is of supreme importance for picking up information about the identity and status of animals both within and outside the group. Female ring-tails perform hand-stands designed to bring their rear ends into comfortable contact with a slim sapling which is assiduously rubbed with genital secretions. The males do likewise, but also grasp a sapling (often the same one that has just been thoroughly smeared by a succession of females) between all fours and, with a peculiar tugging action, mark the wood with a wrist-mounted spur which is impregnated with their scent. The action may be so vigorously performed as to score the bark and expose the sapwood beneath. Both sexes will often form a queue to add their contribution to a particularly important scent post, such as a sapling in the middle of the overlap area which has been previously smothered in scent by an opposing group. In contrast, however, such 'foreign' marks at the periphery of this overlap area are carefully inspected by an inquisitive nose but then left untouched. Presumably at this outer limit 'ownership' by the rival group is tactfully accepted as their due right.

Male ring-tails also employ a long-range scent weapon during inter-male

Even adult primates indulge in light-hearted play sessions. These Verreaux's sifakas (*Propithecus verreauxi verreauxi*) in Madagascar's Berenty reserve ended their long midday siesta by playing 'footsie' and boisterously wrestling before they resumed their foraging activities in the cooler part of the afternoon.

Milne-Edwards diademed sifakas (*Propithecus diadema edwardsi*) often have specific regularly used 'playgrounds' on the forest floor, where all the members of the group come down to enjoy a rough and tumble together on the ground. They also play up in the trees, hanging from a branch and perhaps pulling a companion's tail. Photographed in the rainforest at Ranomafana.

border disputes, wafting 'scent bombs' at one another by vigorously shaking their conspicuously striped tails over their backs in a powerful display which embodies both olfactory and visual components. The male preloads his tail with scent by drawing it through his wrist glands, which heavily impregnate it with odorous secretions. At the same time as shivering his tail and showering his opponent with scent the male also does a handstand and rubs his rear end against a handy sapling, thereby sending out messages at both ends simultaneously. These so-called 'stink fights' mainly take place during cross-border encounters between neighbouring groups. During these it is, as usual, the females who take the lead: they try to intimidate the opposing females by leaping and darting forwards at them, backed up by furious scent marking on

The red-fronted lemur (*Eulemur fulvus rufus*) has been described as a social enigma, with no apparent clear-cut social rules. Unlike the situation in most lemur societies the females are not dominant, although it does seem that it is a female who usually initiates group movements. Photographed in rainforest at Ranomafana, where densities are much lower than in certain areas of gallery forest in the south-west of Madagascar.

any available trees. The males generally eschew this physical stuff and rely instead on their long-range stink weapons.

Social arrangements in some other lemurs do not seem to be nearly as clear-cut and well defined as in the ring-tails. The red-fronted lemur has been described as a 'social enigma' in which any kind of dominance hierarchy seems to be lacking. There does not even appear to be an unequivocal group leader, although it is almost always a female who initiates group movements around the territory. Beyond that, however, females do not seem to be at all dominant – a rare circumstance in lemur society – and usually take second place to males in things which really matter such as access to a particularly rich fruiting tree.

However, females are not always born losers and can on occasion displace a male from a favourite spot; there are simply no strict ground rules in this animal's behaviour. Interestingly enough, in well-habituated groups of both the red-fronted and white-fronted lemur it is always the females who are more daring in making a really close approach to humans and the males who exhibit signs of nervousness, being the first to back off from any unexpected movement. Allegiances within the group seem to be constantly up for grabs and so unstable that they may change back and forth within hours; the red-fronteds studied by Deborah Overdorff at Ranomafana were described by her as 'behaving like a group of teenage girls'. In the crowned lemur females win disputes most of the time, but males are not so thoroughly brow-beaten that they are unable to win now and again – unlike in ring-tails, where the uncompromisingly acquiescent males never seem to fight back. Crowned lemur groups seem to be remarkably loose affairs with no definite dominance hierarchy, although a female always seems to lead movements and both sexes scent-mark the territory as they go, using the 'standard' ano-genital application to rocks and trees.

Multi-male groups are also seen in many species of monkeys and baboons, including the black-capped capuchin, Geoffroy's spider monkey and Rhesus macaque. Howler monkey groups normally consist of one or two adult males along with several adult females, although the males' tolerance of one another may be sufficient to allow up to six to reside within a group. Relationships are hardly electrifying, with relatively little contact between group members – a trait shared with the equally vociferous indri. In these two species, both of which are on the large size for an arboreal primate, extremely loud and far-carrying singing seems to take the place of energy-demanding patrolling or fighting in effecting territorial defence.

As in the indri, howler monkey groups generally alert the broad neighbourhood to their continued presence early in the morning by giving vent to a series of deafening, rather lion-like roars; howls is scarcely the right word, despite their common name. Even so, these impressive post-dawn declarations do not always suffice to prevent neighbouring groups from chancing across one another later in the day. When this happens the scene is set for a formidable and inspiring verbal contest as the males perch high in opposing trees and roar their hearts out at one another across an intervening space of perhaps only 20–30 m (20–30 yds) or so. These oral duels may go on for as much as twenty minutes or more, representing one of the most awe-inspiring phenomena to be found in the South American forests.

In the beautiful red howler, however, there is one context in which verbal abuse serves no purpose – when several males from a rival group gang up to try and take over another group's females by kicking out the males. During such invasions the generally placid demeanour of the average howler is completely cast off and a savage bout of fighting may result in severe injury or even death. Just who wins generally seems to depend on the extent of mutual cooperation between either the invading or defending males: aggressive or defensive pacts between males on either side are seemingly vital in deciding the outcome. If successful, the raiders may consolidate their newly won position of sexual dominance over the group's females by killing one or more of the vanquished males' offspring. Such instances of infanticide are probably far more common among primates than had been thought, even being recorded in such generally peaceful animals as Milne-Edwards diademed sifaka, and it may be

The muriqui (*Brachyteles arachnoides*), largest of the South American primates, seems to enjoy a very peaceable and well-ordered lifestyle in which male–male relationships are remarkably close. Muriqui society is one of the few in which the females emigrate from the group of their birth, rather than the males, the more usual situation in primates. Photographed in Atlantic Coast rainforest at Fazenda Montes Claros in Brazil.

standard procedure under certain circumstances in Hanuman langurs (see pp. 116–17).

In howler monkey society males rank above females and young adults above their elders, who gradually descend the status ladder as they grow older; immature juveniles, however, rank lowest of all. Normally there are few outward signs of this dominance hierarchy except when a subordinate gives way to a superior in matters such as access to food. The muriqui of Brazil's remnant Atlantic coast rainforests – the largest of the South American monkeys – also seems to reinforce the view that large size goes hand in hand with a placid lifestyle in arboreal primates for which leaves are the main form of sustenance. Muriqui groups may contain up to thirty-four members of both sexes. The males seem to get on particularly well with one another – any one male seldom allows himself to drift away from another by more than 5 m (5 yds) or so, and male–male relations are closer in this species than in any other primate. In fact a general feeling of harmony seems to pervade the entire group and belligerence of any kind is extremely rare, even when access to drinking or feeding sites is at stake. Even lengthy mating sessions (up to eighteen minutes at a time) between group members seem to generate none of the annoying curiosity or even outright interference which can make the love life of some other primates, especially baboons, so vexacious and difficult to consummate in peace. This is not the only unusual aspect of muriqui society, for it is the females who habitually leave the group of their birth and join another, leaving the males behind to constitute the stable nucleus of the group. Such a system is rare in primates; a further example is the red colobus, although in that species the males also show some roving tendencies. Female emigration is also found in the gorilla: a female may move home of her own accord, lured away when a neighbouring group comes close enough to effect the switch. She may possibly move on again later, trying several groups before finally settling down. In chimpanzee society, however, the females are generally coerced into leaving the group by pressure from other members. Generally, though, in most primate societies it is the males who are the emigrants.

In some multi-male societies the dominant males employ eye-catchingly explicit methods of demonstrating their superiority to a subordinate without having to prove the point by picking a fight. In the savannah-living vervet monkey the alpha male flaunts his genital region as a kind of 'badge of office'. In this species the scrotum is a conspicuous powder blue, against which the scarlet penis shows up particularly well when fully extended. To complement this vivid display there is a patch of red skin at the base of the tail, and below that a patch of white fur. A high-ranking male of a mind to reduce a subordinate to a state of cringing submission needs merely to stand up high on his rearlegs and strut in a cocksure manner around the inferior, who adopts a suitably hunched posture in response. As the alpha male circles, the grovelling object of his visual put-down also turns, but at a slower rate, constantly keeping the brilliant genital display in sight. The subordinate also keeps his own genitals well out of harm's way by retracting the scrotum into his body and keeping his rear end pointed well away from his circling tormentor. This protective attitude towards his vulnerable genitals may be crucial, as alpha males have a habit of eliminating once and for all a subordinate's future prospects as a sexual competitor by dashing in and taking a quick and conclusive bite at a carelessly exposed scrotum.

This little drama is the so-called 'red, white and blue display', which has been seen being given (with various small differences in detail) by vervets over a wide area. For example in Natal the displayer holds his tail in a curve in order to avoid spoiling the impact by covering his genitals; in Kenya the tail is held erect like a pleased cat. There are also variations from area to area, in the actual mode of giving the display, indicating that social behaviour in the vervet, as in most primates, is not set in tablets of stone but is able to respond to local conditions. During the red, white and blue display the recipient is normally thoroughly intimidated and cringes, giving vent to submissive wails and casting rapid nervous glances at the strutting performer. The latter uses his performance to reinforce his right of access to the females, a vital consideration in any multi-male group in which constant militant competition for mating rights could be potentially lethal for the participants and destructive of group cohesion. Indeed, when the females actually come into oestrus mere visual exhibitions alone may no longer be powerful enough to prove superiority. When this happens a real and very nasty fight may be the only way of sorting out the winners and losers; it can result in some unpleasant injuries, especially to such vital equipment as the genitals.

In wedge-capped capuchins (*Cebus olivaceus*) the dominance hierarchy among the females can lead to a rather unusual technique for 'pulling rank' by some of the more dominant individuals: parasitic nursing behaviour. Infant primates are often born hustlers and will routinely try and hijack a meal of milk from a 'foreign' nipple if one happens to be handy and there is a chance of getting away with it, and there is often a good chance of that. However, in the wedge-capped capuchin it is high-ranking adult females who poach milk under duress from their low-ranking companions. The latter do not appear to take kindly to such an infringement of their privacy and exhibit distinct signs of distress, but they have to sit and suffer as they are in no position to do anything about the situation without risking punishment, though thankfully such forced nursing episodes are normally short-lived. The reason for this strange behaviour is unclear; it may simply be a rare but regularly practised method of emphasizing status among the females.

The desire to avoid retaliation when approaching a superior has led to the evolution of a peculiar kind of behaviour in both barbary macaques and savannah baboons. In most primate societies the young juveniles are sacrosanct and can behave in the most outrageous way with even the most senior members of the group without fear of reprisal. A special spot of colour or brightly coloured pelage may be the key which switches on this tolerant response; or, more likely, switches off the normal aggressive reaction. Males of three subspecies of *Papio cynocephalus* have frequently been observed using infants as 'passports' to approach another male, often one in consort with a female, without fear of being beaten up. The infant is treated casually or even quite roughly by its kidnapper, being picked up and treated as a necessary object with a useful role to play in allowing a close approach in safety. The precise reason for this infant-napping is difficult to establish, as the guilty party may use his reluctant safe-conduct pass merely to approach the superior male and not make any physical contact.

Similar behaviour is seen in the barbary macaque, although in this species the young are quite closely involved with the adult males, even to the extent of sometimes being given lifts on their backs. However, when exploited as an aggression deflector a young juvenile in his clutches allows a subordinate male to approach a superior without risk of attack, and such baby borrowing is quite common in this species.

With few exceptions an infant is generally only treated with tolerance by males within its own group. One or more of these will probably have fathered all the resident juveniles, so showing tolerance makes good sense. However, there are records of infanticide within a group, although the circumstances which bring this about are seldom clear. For example, up to 1984 infanticide had been observed in chimpanzees once in the group at Budongo Forest, Uganda, seven times at Gombe Stream in Tanzania and six times at Mahale National Park in the same country. These episodes fall into three categories: the killing of an infant in one group by males from another; the killing of an infant (almost always a male) by males of its own group; and the killing of an infant by a female of its own group. Just why males should kill infants in their own group is a mystery, as chimpanzees' social arrangements decree that all or most of the infants within a group will have been sired internally. This suggests that the murderers are almost certainly killing their own offspring, a most peculiar act from a genetic point of view. In one case at Mahale, however, the attacking males may have been under the impression that the slaughtered infant had been sired by a male from another group, as the mother had been absent for a time, but observers' records suggest that this was incorrect and the most likely father was among the killers. The actual act in this instance was rather gruesome, and the mother suffered a considerable mauling in her efforts to protect her offspring by using her body as a shield. In most cases of chimpanzee infanticide the triumphant killers end up by eating their grisly trophy. In one well-observed instance the victim's brother begged for and received some of the meat from his brother's corpse – just part of a general feast in which several females also enjoyed various bloody titbits from the still warm body of the murdered infant.

Infanticide in species exhibiting the 'one-male' form of social system is perhaps a little easier to explain. In one-male groups a single male tries to defend access to a varying number of females against invading males from other groups intent on supplanting him. This form of society is found in species as varied as Geoffroy's spider monkey, agile mangaby, guereza and gorilla. In the widespread and highly successful Hanuman langur, one-male groups occur alongside multi-male and all-male groups. In multi-male outfits there may be a gradual and peaceful leadership succession which does not involve the death of any of the group members. By contrast, lone males trying to defend their exclusive property may enjoy little success against the kind of determined cooperative offensive which seems typical of this species. In one encounter which was closely watched a group of males attacked another group containing a single male accompanied by nearly thirty females and immatures. Despite their best efforts to defend their babies three females lost their offspring to the invading males, one of whom subsequently supplanted the resident male who had been ineffectual in countering such a concerted onslaught. Bereft of their progeny the bereaved mothers soon offered themselves to their new master and mated, giving birth to his issue a few months later. To us the females' actions may seem incredibly heartless – but after all, they do not want a cissy for a mate who may waste their reproductive potential by fathering a succession of equally gutless offspring lacking the ruthless determination necessary to depose a group leader and sire a line of offspring for themselves. So it is logical to make overtures to a male who exhibits all the signs of being one of life's winners.

The peace of Hanuman or grey langur society may be broken every two to three years by the removal of their leader by a band of aggressive males, one of whom will become the new leader. Both during and after the takeover any babies within the group are at great risk, for infanticide seems to be relatively common in this species as a method of accelerating female sexual receptivity to the new male.

The new leader normally harasses some of the other infant-bearing females soon after his conquest, but his baby-killing endeavours require skill and cunning as he has to reckon with the rest of the females who will often gang up to beat him off. Such attempts to weed out a predecessor's offspring after a takeover seem to be more common than pre-takeover infanticide. However, the latter situation has a greater chance of success if several males act in consort to overwhelm the opposition. This makes it easier to track down the mothers, who normally quickly see what is coming and do their best to hide away from the would-be killers of their babies. Some of the older females may have seen it all before and know the mortal danger to their offspring posed by the invading band of males. After all, their current lord and master probably won his job in the same way, and their new leader can probably only look forward to a tenure of around two years before he in turn is toppled in a cycle of takeovers which seems to typify Hanuman langur society.

The reasons for such acts of infanticide have been disputed, but seem to involve an acceleration of the coming into oestrus of the bereaved females. They would otherwise not be sexually receptive while still nursing babies which to the new master are not only superfluous to requirements but potential competitors. Similar takeovers accompanied by infanticides also occur in several other monkey species, including blue- and red-tailed monkeys in Africa and silvered and purple-faced langurs in Asia.

Hanuman langur society eventually settles down after such a traumatic episode and apart from this the females lead very stable lives, forming the nucleus of the group over long periods. Although there is an age-related dominance hierarchy among them (young mature animals gain dominance over their older colleagues, and then drop down through the hierarchy as they too age) they spend a lot of time grooming one another in matronly harmony. This friendly accord also extends to the great indulgence shown towards handling one another's young, even when newly born. This is a common trait among the colobines and quite the opposite of most cercopithecine mothers, who exhibit extreme protectiveness of their offspring until they reach quite an advanced age.

Gorilla family life generally seems to be very amicable: the silverback group leader may groom his females as well as enjoying their attentions. He also shows extreme tolerance towards the youngsters in his group as they playfully clamber over his body. In fact his giant figure seems to be a positive attraction for juveniles of varying ages, who seem irresistibly drawn towards him in a bid to sit near him or even groom him. On the other hand, lone males intent on wife theft will bulldoze their way into a group and abduct a female after first getting her infant out of the way by killing it.

Patas monkey males seem to lead rather lonely lives, despite being the head of a group of females and offspring. The females appear to be notably anti-male for much of the year, actually threatening their leader if he comes close. They only reveal any clear signs of friendliness when they come into oestrus, when it is they who solicit his attentions. The group leader's wives, by contrast, lead very gregarious and affable lives featuring long grooming sessions and a great deal of mutual baby-sitting. Patas males too form single-sex groups in which they enjoy a much richer social life, forming close bonds with certain selected companions. What a marked difference from the isolated life of the leader in a mixed group, who seems to pay the price for sexual success by becoming almost a social outcast. Patas monkeys roam over enormous ranges in their dry savannah environment – up to 38.5 sq km (15 sq miles) has been reported – and when neighbouring groups make the occasional chance meeting they try to avoid any contact. The only exception happens during the breeding season, when rival males may raise the temperature somewhat by making aggressive chases and postures towards the other side.

Chapter 6

Food and Feeding

Many primates owe their success to their ability to find and digest a wide variety of different foods, while others are more specialized and are adapted to cope with a more restricted diet such as leaves. Even when a single class of food, such as leaves, figures prominently in the diet at one season it may not be so important a few months later, when fruit may be the main item on the menu. Meat is eaten by more species of primates than was once believed, and the recent increase in field observations has made it clear how regularly insects, lizards and even small birds and snakes are eagerly taken by primates hitherto considered to be mainly vegetarian.

MEAT

This kind of food comes in two main forms: hard and crunchy on the outside and soft and tasty on the inside, as in millipedes and such insects as beetles, bugs, crickets and katydids; and soft on the outside and bony within, as in frogs, lizards, snakes, birds and mammals. Insects form the major part of the diet for a number of primates, notably several of the prosimians, as well as adding variety to the staple fare of many others. Since they are invertebrates insects wear their skeleton on the outside – hence the crunchy nature of the exterior, which forms a tough shell protecting the vulnerable vital organs. This so-called exoskeleton is composed of a highly weatherproof and durable substance called chitin, notable for its indigestibility although it is potentially nutritious. Because of their generally unrewarding nature the highly chitinous parts of insects such as the legs and wing-cases are often discarded by insectivorous animals, including primates.

The chief exponents of the carnivorous lifestyle are the tarsiers, which are the only primates with exclusively predatory habits – they never take any kind of plant food. They are superbly equipped for their role as night-hunters, their enormous goggling eyes acting in perfect consort with their highly sensitive ears to detect the slightest movement by a potential meal. Even if the soft pattering of a beetle's legs on a broad forest leaf or the buzzing song of a love-struck katydid comes from behind the questing tarsier it doesn't matter, for with its flexible neck the tarsier merely needs to swivel its head round to look back over its shoulder in the uncanny way so familiar in owls. A quick bound, and the prey is instantly despatched in a flurry of snapping teeth. Such is the precision with which the fatal bite is administered that tarsiers do not hesitate to take on small poisonous snakes, which are given few opportunities to retaliate with their deadly strike. The same goes for scorpions which are

Tarsiers will tackle even such large heavily armoured insects as this rhinoceros beetle, although the horns and spiny legs may be capable of inflicting damage on the tarsier's face and hands before success is assured.

given little chance to swing into action with their pivoted tail-mounted sting and are dealt with as speedily and efficiently as the harmless cockroaches which form an important part of a tarsier's diet.

Subduing some of the heavyweights of the insect world can, however, prove a tough challenge for a small animal like a tarsier, for which over-confidence can prove awkward. An observer was on hand to watch the fascinating struggle between a particularly determined Horsefield's tarsier and an extremely large and well-armoured rhinoceros beetle. The tarsier's sharp teeth proved inadequate to the task and it took fifteen minutes to broach the tough, smooth armour-plate protecting the beetle's thorax. It finally cut short the insect's struggles, but not before this battle-tank of a beetle had managed to get a purchase with its strongly projecting horns and inflict a certain amount of damage on the soft skin of its furry tormentor. Despite enduring such ignominious treatment by its prospective meal the tarsier was game to the end and showed no signs of wanting to give up and look for an easier target. This is perhaps not surprising, as a rhinoceros beetle makes a substantial meal and the scarcity of such fat rewards in the rainforest dictates that when found they should not be easily relinquished. So the tarsier persevered, licked its wounds and ate the beetle. Small birds and bats succumb in a similar and perhaps less belligerent manner, while stick and leaf insects (phasmids) are easy game, for even their incredibly faithful mimicry of items of vegetation, which do not normally feature on a hungry tarsier's hit-list, probably avails them little in competition with a tarsier's ultra-sensitive nose and ears.

Similar hunting techniques using sophisticated eye–ear–nose coordination

When the mohol bushbaby (*Galago moholi*), pictured here, coexists with another species such as the thick-tailed bushbaby competition is avoided because each species targets different insects.

are employed by the African bushbabies. As they scurry through the complex network of branches comprising their arboreal territories bushbabies probably provoke resting insects into making sudden moves which betray their presence. When confronted with an insect making a precipitate departure, such is a bushbaby's agility and speed of reaction that it is able to employ a grab-and-smash technique to cut short the prey's escape, deftly snatching the insect out of the air before a quick bite puts an end to further movement.

Where more than one species of bushbaby occupies the same habitat a certain amount of wasteful competition is avoided because each species targets slightly different kinds of insects. In a forest in South Africa where the mohol and thick-tailed bushbabies occur together the former was found to take a greater proportion of beetles, dragonflies and antlions, while the latter took far more ants and also tackled fair quantities of centipedes and termites, which were not recorded in the mohol's diet. Both species took approximately equal quantities of grasshoppers and katydids, probably reflecting these insects' generally high acceptability rating to many predators coupled with their relative ease of capture. Whereas the mohol devoted much of its time throughout the year to searching for invertebrates, its larger relative generally made this more of a summer task, switching to alternative food supplies such as gum in winter, although gum featured quite prominently in the diet of both species throughout the year. In a Kenyan site the food of *Galago zanzibaricus* was found to comprise around 70 per cent insects, most of which were lumbering, easily caught beetles. Strong chitin-crunching teeth rather than lightning reflexes would be the most important tools for coping with such a diet, which is supple-

Many bushbabies include varying amounts of gum in their diet, mainly derived from various species of acacias. In this respect they are similar to the tree or acacia rat (*Thallomys paedulcus*), which gnaws regularly visited 'sap mines' on acacia bark, often living in the same areas as bushbabies.

mented with fruit. In the same habitat the slightly larger *G. garnettii* munched its way through more or less equal amounts of insects and fruit, but diversified from this into a role as big-game hunter by taking on amazingly outsize birds such as the chicken-sized crested guineafowl, which were reduced to scraps of feather and feet by this miniature but surprisingly fierce giant-killer.

Members of the Lorisinae also prey heavily on insects, not in the energetic, agile manner seen in tarsiers and bushbabies but in a slower, more methodical fashion more suited to their rather lethargic approach to life in general. Creeping forward chameleon-like with measured step, a slender loris sniffs its way through the darkened forest tangles, the very epitome of the stealthy hunter – although not lacking in speed and precision when the need to avoid losing an insect means gathering it out of the air with a firm, two-handed grip. Snails oozing their way quietly across a forest leaf require less finesse and are noisily pulverized between powerful teeth, and on occasion the odd lizard, bird's egg or the bird itself, caught unawares whilst peacefully roosting, all add bulk as well as quality to the loris's run-of-the-mill invertebrate fare.

Along with the slender loris those other members of the Lorisinae, the slow-loris, potto and angwantibo, together form a select band of hardy and enterprising gourmets which seem to be both indifferent and resistant to the chemical defences deployed by certain insects so successfully against other types of predators such as birds.

Many plants attempt to defend their leaves against herbivorous animals by lacing them with powerful doses of chemicals. Such substances are often successful deterrents against attack by vertebrate predators such as cattle or most leaf-eating primates; much less so against insects. In fact the defensive chemicals in many plants have actually come to function as feeding *attractants* rather than repellents for many insects, particularly in their larval stages which sequester the chemicals in their bodies in order to redeploy them in a highly concentrated or perhaps suitably modified form (making them even nastier) in their own defence.

Many of these insects protect themselves from day-active, visually hunting predators by flaunting bright, easily memorized warning colours. Being nocturnal, the members of the Lorisinae cannot appreciate such flamboyant 'keep off' signs, but the utterly repulsive chemical aura surrounding such insects would, one might think, convey an equally persuasive negative message to their sensitive noses. Far from it. By their very nature these nauseatingly smelly insects tend to be slow-moving, relying on their bright uniform and general nastiness to deflect an attack rather than resorting to fleetness of foot or wing. This works well against most enemies but not against the members of the Lorisinae, which seem to be made of pretty strong stuff, loath to forfeit such easily caught prey no matter how revolting the stench or disgusting the taste. However, some of these insects also pack quite a toxic punch, since they contain powerful poisons capable of causing severe illness in any vertebrate unwise enough to eat them. Vomiting may be the least of the consumer's worries; fatal cardiac arrest could follow. But the Lorisinae, it seems, have few such problems for they belong to a relatively small and exclusive group of predators which seem immune not only to their prey's objectionable flavour but also to its poison. This places the Lorisinae in a unique position to capitalize on easily detected and lethargic targets which would be left severely alone by most self-respecting insectivores.

Although considerations of poisoning do not allow the average mammal to make an easy meal of chemically protected invertebrates, such obstacles are not necessarily insurmountable for animals as intelligent as primates. At least one species has devised a neat way of converting an otherwise disgusting and inedible food item into an acceptable if rather crunchy and unappetizing meal.

In the eastern rainforests of Madagascar the red-fronted lemur and red-bellied lemur frequently chance upon several species of giant millipedes. These fall into two categories: slow-moving *Sphaerotherium* spp. pill millipedes, which when disturbed roll up tightly into defensive spheres the size of golf balls; and 20 cm (8 in) long juliids, which wend their way across mossy logs and branches on smoothly flowing legions of busy feet. Both kinds are covered in shiny black or brown armour which is often emblazoned with bright orange 'warning' stripes or spots; in other species rows of bright yellow legs contrast with polished black bodies. This 'warning' livery is supposed to prevent enemies from even touching them, and indeed the forest-wise lemurs seem to know all about the millipedes' caustic secretions which ooze forth to form a protective coating like a toxic dew glistening on the millipedes' smooth exteriors. They are only produced, however, when the millipede is picked up and roughly

Despite its warning uniform this 15 cm (6 in) long giant millipede falls prey to red-fronted (*Eulemur fulvus rufus*) and red-bellied (*E. rubriventer*) lemurs in the forests of Ranomafana. The millipede's repellent secretions are rubbed off by the lemurs during a lengthy session of preparation before the meal commences.

handled, so the lemurs seem deliberately to provoke the copious production of these secretions by rolling the millipedes between their hands. Now and again they vary this pre-treatment by drooling over their victim and bathing it with saliva before going on with the hand-swabbing routine, occasionally swapping over to using the tail, which they hold like a dishcloth between their hands. It is possible that something in the lemurs' saliva may neutralize the toxic quinones discharged by the millipedes, while the physical rubbing must remove the bulk of these offensive substances. Either way, once their careful and lengthy preparations are complete the lemurs tuck into their meal, biting noisily through the hard armour of the tightly rolled-up pill millipedes like a child with a crisp green apple.

Lemurs are not generally reckoned to be the eggheads of the primate world but these two species seem to have figured out the solution to this particular problem, although it may originally have come about by accident. It would be interesting to know whether both species solved the problem independently or whether one came up with the answer first and was copied by the other. Such apparently sophisticated meat-eating behaviour is rather surprising in two lemurs whose main diet consists of fruits, flowers and leaves, and neither of which had formerly been considered as predatory in the wild state.

The two *Microcebus* mouse lemurs are among the smallest of primates, yet their diminutive stature is coupled with a fierce nature which enables them to tackle surprisingly large prey such as chameleons (which abound in Madagascar) and birds. These together form part of an amazingly varied diet which includes almost anything else which is edible – such as tree sap, insects, spiders, tree frogs, birds' eggs, flowers, fruits and even a few leaves. With such a catholic diet allied to their small size, which enables them to thrive in forest remnants, it is perhaps not surprising that these are the least threatened of all the lemurs.

Their close relative, Coquerel's mouse lemur, is rather larger. During the bountiful nights of the wet season it enjoys a similar diet, but during the long dry season things begin to get tough as the forest's food supply dwindles. It is during such hard times that *Mirza* falls back on one of the most unusual resources utilized by any primate. Large, densely packed colonies of flatid bug nymphs adorn the leafless branches of the trees during the height of the winter drought. These insects continuously suck away at the branches with such unstinting dedication that the throughput of sap is considerable. Much of this passes straight through and is vented as a sweet, sticky honeydew, a waste product as far as the bugs are concerned but a vital and greatly relished resource for Coquerel's mouse lemur. In fact at this season it devotes 60 per cent of its feeding time to lapping up these sweet exudates. Its success is increased by the bugs' helpful habit of squirting out a few extra drops of fresh honeydew when disturbed by the lemur's nightly activities, a kind of 'service on demand' which must greatly increase the yield and therefore the usefulness of the supply.

The exact nature of the diet taken by the weirdest of the lemurs, the aye-aye, has been the subject of some debate in recent times. It has often been called the Madagascan equivalent of the woodpeckers, familiar wood-probing insectivorous birds which are absent from the island. Aye-ayes go about the business of obtaining a meal using methods which are highly unconventional by any standards. For a start they are not known to take surface-living insects, and unlike bushbabies and lorises an aye-aye will probably walk straight past a tasty grasshopper perched invitingly on a leaf or tree trunk. The aye-aye is more interested in what lies hidden beneath the bark, particularly inside the many lianas as thick as a man's arm which drape the trees in its rainforest home.

The animal's searching technique is well developed and highly efficient. Inching its way up the swaying stem the aye-aye focuses its huge bat-like ears forwards, listening intently for the faint sounds of a beetle larva rasping away at the wood beneath. If silence reigns beneath its feet the aye-aye may tap a rapid tattoo on the bark with its skeletal third finger, moving up the liana and tapping as it goes. Why it does this is not clear; it may reveal hollow insect galleries by a kind of echo-sounding technique, or perhaps it induces the hidden larvae to produce some kind of sound which gives them away. Once something of interest is located the aye-aye gets down to business with a vengeance: now the reason for its ever-growing, beaver-like teeth becomes obvious as chips of timber are ripped away and flung in all directions, pattering noisily to the forest floor. Such is the din made by an aye-aye wreaking destruction in this manner that listening out for it is one of the best ways of locating aye-ayes in an area of forest.

With its quarry uncovered, the aye-aye desists from its demolition tactics and resorts to more finesse, delicately probing into the exposed gallery with its skeletal third finger. Some observers have noted that this is stirred around in order to convert the hapless beetle larva into a pulp which is then licked off the finger; other observers have noted larvae being extracted unpulped, skewered on the tip of the finger like soft white kebabs, to be consumed with gusto with a few snaps of the aye-aye's fearsome teeth. Aye-ayes will even descend to the ground to cause havoc on fallen logs which are in an advanced state of decay. By this late stage the solid timber may have been converted to no more than a kind of soggy sponge which can be pulled out in handfuls; the aye-aye's teeth thus have no problems in quickly uncovering large longhorn beetle larvae quietly feeding within their mouldering home which disintegrates around them, leaving them plump and helpless, to be impaled and eaten.

The aye-aye's chisel-like teeth and skewer-like third finger are also used to exploit another highly unusual food source. Near the shore on the island of Nosy Mangabe grow numerous *Afzelia bijuga* trees whose trunks are scarred by large, woody galls. At certain periods of the year the aye-ayes devote more of their foraging time to these excrescences than to anything else, biting into them to expose a yellowish, clay-like tissue. This has quite a strong, sweet smell and seems to be greatly relished by the aye-aye, which will spend a considerable time dipping its fingers into the tissues, extracting a few crumbs which are licked off the finger like a child with a sherbet dip. Two types of gall occur, one strongly projecting and knob-like, the other somewhat star-shaped and more flattened. Some individual galls seem to be more attractive than others, being visited night after night so that the aye-aye's repeated gnawings create a hole in the bark. These favourite feeding sites can be easily recognized by the characteristic grooves made by the aye-aye's teeth; yet nearby galls may be completely ignored. The edible portion of the galls obviously penetrates to a depth inaccessible by gnawing alone – hence the finger-probing needed to extract as much as possible.

The galls are probably induced by a fungus able to break down the cellulose content of the wood and convert enough of it to sugars and other digestible products to make it a nutritious food for the aye-aye, especially when insect larvae are in seasonally short supply. Many of the galls also contain click-beetle larvae (Elateridae) which provide a bonus for the aye-ayes, although these grubs are probably secondary residents and not the main attraction.

Many kinds of monkeys, of both New and Old World types, include insects and small vertebrates in their diet – some as an integral and regular part, others on an opportunistic basis when chance permits. In the Amazonian rainforests the saddle-backed tamarin may spend as much as 45 per cent of its active day looking for insects and other animal prey, mainly small tree frogs and lizards such as anoles. This tamarin prefers tree-fall habitat with tangles of exuberant new growth – just the place to find plenty of insects, for such 'edge' habitats always attract a proliferation of invertebrate life. Fruit is the most important plant food for the tamarins, with nectar and plant exudates following far behind, except in the dry season when the latter attain more importance. Such a catholic diet is typical of the callitrichids as a group, although the importance of insects versus gum or fruits and so on varies from species to species; mature leaves never seem to be eaten.

Although squirrel monkeys eat considerable quantities of flowers and fruits, their preferred food seems to be insects and spiders (during seasonal shortages of fruit they will become exclusively insectivorous) with larger vertebrates such as small frogs and lizards and the odd bird's egg thrown in when available. A group of squirrel monkeys ransacking the complex tangle of the rainforest canopy in a painstaking search for the multitude of insects and spiders skulking within is rather like a Customs rummage squad going over a tourist bus suspected of smuggling drugs. The cavernous leaf bases of a bromeliad rosette perched high on a branch provide refuge for a host of insects, so the wise squirrel monkey devotes considerable care to rifling through the collection of dead leaves and other refuse which accumulates there. Every shrivelled leaf will be carefully inspected and turned over to make sure that nothing is hiding in an unseen nook or cranny.

One suspects that such careful scrutiny is also born out of the experience that some of the 'leaves', both living and dead, are not quite what they appear to be. Many tropical katydids (Orthoptera: Tettigoniidae) and praying mantises (Mantodea) mimic leaves with astounding verisimilitude; it therefore pays a hungry monkey to be certain that what is being handled really is a useless dead leaf and not a tasty katydid about to sprout six legs and make a quick getaway. This is just what does happen when a squirrel monkey troop is

Many tropical bush crickets or katydids (*Tettigoniidae*), such as this Venezuelan *Typophyllum* sp., bear a remarkable resemblance to leaves, both living and dead. It is not surprising, therefore, that resourceful primates such as squirrel monkeys carefully turn over and examine dead leaves before rejecting them.

Many insects such as this Costa Rican 'peanut bug' (*Laternaria laternaria*) employ a defensive display embodying the sudden exposure of previously concealed eyespots, usually on the hind-wings. This may induce the attacker, such as a squirrel monkey, to release the insect in surprise, and indeed monkeys have been seen jumping backwards with fright when abruptly confronted by such an unexpected 'face' staring at them.

Members of the Lorisinae seem capable of eating some remarkably unpleasant chemically defended insects. However, it is rather surprising that the red-backed squirrel monkey (*Saimiri oerstedii*) in Costa Rica has been seen stuffing down mouthfuls of these small, warningly coloured nymphs of the giant grasshopper *Tropidacris cristatus*.

brashly crashing about in the canopy – there will be a hail of panicked katydids and other insects plopping on to the forest floor, having taken the quickest way out by diving earthwards. However, the monkeys are often too quick for them and snatch a great deal of their prey off leaf surfaces.

Large caterpillars are sitting targets and a plum prize, even causing one of the rare disputes over ownership. In Costa Rica *Saimiri oerstedii* have been seen grabbing handfuls of *Tropidacris cristatus* grasshopper nymphs and stuffing them into their mouths like a child with popcorn. These nymphs live in dense bands, so would seem to be easy meat. However, the nymphs' aggregating habits are not intended to make them easier targets for marauding squirrel monkeys but to emphasize the warning pattern which each nymph carries. This should theoretically protect them against visually hunting predators – but nobody told the monkeys, who chew away on them by the mouthful with no obvious immediate signs of disgust nor any subsequent inclination to vomit their colourful meal back up again.

The African equivalent of a group of foraging squirrel monkeys is a mangabey troop setting about its daily task of rifling through the forest. Mangabeys are considerably larger and heavier than squirrel monkeys, so can do an even more thorough demolition job on their surroundings, ripping of bark with their powerful teeth in a quest for insects and spiders beneath, and terrorizing nesting birds in pursuit of the considerable bonanza of a whole clutch of eggs or nestlings. However, such encounters do not always go the monkey's way; fearless and persistent mobbing of their tormentor by the parent birds may succeed not only in driving the intruder from their nest but in sending it cringing under cover until the dive-bombing has ceased. Like many primates, however, the mangabeys' main food is fruit and flowers; leaves do not figure largely in their diet as their digestive systems are inadequate for dealing with large quantities of cellulose. Unlike the squirrel monkey a mangabey can extract maximum reward from a rich but limited food source by stuffing full its capacious cheek pouches and then continuing its meal at leisure later on, when the rest of the group cannot cash in on the discovery.

It was once thought that chimpanzees were almost exclusively vegetarian, feeding on a very wide range of plant species. Now it is being increasingly realized that fresh meat is a greatly prized element of the diet in at least some groups, although there seems to be considerable variation in actual hunting techniques and in the degree of consumption and type of prey taken by different populations from diverse areas. At Mount Assirik in Senegal, for instance, chimpanzees raid the honey stores of wild honey bees and do not neglect to eat the well-armed owners as well. Other insects with a proven ability to strike back at their tormentors, such as *Camponotus* spp. and *Megaponera foetens* ants, are also eaten. The nests of the tailor ant (*Oecophylla longinoda*) also provide a convenient source of protein. This green living tent, consisting of a few leaves fastened together by silk, forms a handy edible sachet well crammed with a tasty filling consisting of a sizeable number of eggs, larvae and pupae as well as adult tailor ants. The chimpanzees treat these nests as fast food, plucking them by hand and eating the lot, leaves and all.

The large soldiers of the African driver ant (*Dorylus nigricans*) are equipped with formidable jaws which can inflict a painful bite. Chimpanzees have varying methods of digging or 'fishing' for these aggressive insects which minimize the amount of painful retaliation received.

The glossy black driver ants (*Dorylus nigricans*), with their legions of sickle-jawed soldiers ever eager to pounce on an intruder and administer instant painful deterrence, would not at first sight appear to offer much gastronomic promise to even the hungriest of chimpanzees. Yet these fierce mini-warriors, which terrorize all and sundry as they rampage through the forest in search of prey, do not appear to intimidate at least some chimpanzees. The author's (KGP-M) first encounter with these tiny pugilists led to his jeans hitting the ground in double-quick time in an effort to discover who seemed to be sticking lighted matches into his stomach. Yielding flesh being thoroughly chewed by several soldier driver ants revealed the cause, while the sight of two columns of reinforcements coming to their aid up each boot sent the author hopping, jeans around ankles, out of the danger zone. Chimpanzees, however, regularly seem to be willing to suffer such painful consequences in return for a few mouthfuls of these pugnacious horrors.

In the Tai National Park in the Ivory Coast the plunder of the ants' subter-ranean fastness involves two different methods. The first is the more rewarding but also the more painful, as the chimpanzee must make a direct physical intrusion into the nest. To gain access the would-be diner must first rake away the loose earth covering the entrance. This is performed with a strong sense of urgency born out of the need to get the job done quickly before armed Nemesis pours forth to take revenge on the perpetrator. Once the nest is open the chimpanzee boldly sticks its arm up to the shoulder into the seething mass of ants, grabbing a fistful of succulent larvae from the lowermost reaches where they are kept for maximum security. However, business must now come before pleasure and the triumphant raider first has to spend a few frantic moments brushing away hordes of furiously biting ants firmly attached up the length of its arm before it can wolf down its booty.

The second method is more sophisticated and less hazardous, needing no bodily intrusion into the nest. This is because it involves a habit much associated with chimpanzees and one which holds a particular fascination for human observers – tool-using. Here the chimpanzee combines the tool's inert nature with the soldier ants' defensive instincts to secure a regular supply of ants in a less painful way. The tool is a stick 20 cm (8 in) or so long which is dipped into the mouth of the nest. The cohorts of soldiers poised ready to repel any such invasion immediately start to stream up the stick in some number, when they are whisked upwards towards the chimpanzee's waiting mouth and swiped dexterously inside by a flick of the lips before they have time to do much about it. Such is the slickness of this operation and so large the reservoir of ants ready to attack the trespassing 'fishing rod' that the chimpanzee is able to dip up a fresh supply of ants as much as twelve times per minute. It seems that around fifteen ants are allowed to swarm up the stick before the waiting chimpanzee decides that they are getting too close to its hand for comfort and swipes them into its mouth. That makes around 190 large, protein-rich ants descending into the chimpanzee's stomach every minute – no wonder a few bites on the hand or lips are considered a small price to pay for such rewards.

At Tai in the Ivory Coast 'fishing' is seldom employed for the large black *Dorylus nigricans* ants, nest plundering being preferred for this species which only has a moderately painful bite. The smaller *D. gerstaeckeri* can inflict more pain but the soldiers are slower-moving than in *D. nigricans* and less likely to reach the hand of a fishing chimpanzee by running up the stick too fast. Therefore 'fishing' is the preferred method for this species, due to the unacceptable degree of punishment inflicted on any arm unwisely delving into their sanctum.

For some reason the skills involved in fishing seem to be more common in the females than the males of the group. The males rarely seem to bother, being more adept at the less subtle strong-arm smash-and-grab attacks on the nests which are seldom practised by the females. It is interesting to compare the frequency and nature of these techniques at Tai with those used by the chimpanzees at the Gombe Stream reserve in Tanzania.

At Gombe the fishing rods are much longer, averaging 66 cm (26 in) in length, enabling large numbers of ants to swarm up their considerable length before the chimpanzee needs to terminate their advance. This is done not by sweeping the rod through the lips but by swiping the ants off in the other hand, which then transfers the struggling mêlée to the mouth. This method is presumably more painful than the Tai chimpanzees' direct technique, as the

Gombe procedure allows the ants a second chance to bite the hand that grabs them before the waiting gourmet's teeth settle the matter. However, the longer Gombe-style stick does have the advantage of landing a greater catch at each dip, with an average of 292 ants being netted each time. Dipping frequencies at Gombe are slower – just under three per minute – but the greater catch each time does mean that nearly 900 ants per minute are taken compared with only around 190 per minute at Tai. Fishing seems to be the preferred method at Gombe, where nest delving is rarely seen.

The chimpanzees at Tai also know how to extract the fat juicy grubs of carpenter bees (*Xylocopa* sp.) from their tunnels in dead wood without being vulnerable to agonizing retribution from the fierce sting of the defending female. The chimpanzee first 'tests the water' by poking a stick into the nest hole to see if anyone is at home, whereupon a resident female will react by shoving her backside into the entrance with her sting at the ready. This is just what the chimp has been waiting for, as the stick makes a handy lance for injuring the bee so badly that she tumbles from her nest and is eaten. With the guardian out of the way the chimpanzee can now tackle the defenceless grub by biting away the wood. Once exposed, both the larva and its store of honey are devoured.

The cache of honey in a carpenter bee's nest is rather meagre and cannot compare with the bountiful reserves in the much larger nests of such social bees as the honeybee (*Apis mellifera*) and some of the stingless or sweat bees (*Trigona* spp.), and considerable risk-taking may be justified to obtain such copious supplies. Chimpanzees seem willing to gamble on receiving a peppering of stings when reaching into a nest to grab as much honeycomb as they can. Next comes a hasty retreat to eat their spoils free from a swarm of enraged bees. Disturbed nests are dealt with more safely by again resorting to fishing methods, a stick being dipped repeatedly into the nest and the honey and wax licked off. Stingless bees are incapable of such painful retaliation, so their nests are directly torn open with the chimpanzee's jaws; tools are only used as back-ups to extract what little honey remains inaccessible to the robber's questing teeth and tongue.

The chimpanzee's most familiar form of fishing for insects invites little fear of annoying retribution. Many African termites (Isoptera) live inside large fortress mounds of hardened earth so that before fishing can commence the chimpanzee has first to broach the outer wall. Termite-fishing behaviour has now been reported from widely scattered localities in Gabon, Senegal, Equatorial Guinea, Cameroun and Tanzania. Several methods are used, mainly varying in the length and diameter of the stick employed as the fishing rod. As with the driver ants it is the soldier termite nest guardians which are the main quarry, swarming hostilely on to the stick before being swept into the mouth. Termite colonies tend to be very enduring affairs which remain in the same place for a very long time. Chimpanzees can therefore depend on them for a regular and predictable source of food, unlike the peripatetic driver ants whose nests are temporary encampments liable to change their position with confusing unpredictability.

'Fishing rods' are usually tools in the true sense of the word, in that the original sticks are suitably modified by the user in a predetermined way in order to carry out a specific task. It is important to stress that the toolmaker foresees the need for modification in advance of the job. Thus sticks for gathering termites, ants or honey may be of different lengths and thicknesses, depending on the requirements of the task in hand. The modification usually consists of stripping the twig of leaves and offshoots and making sure that it is the correct length. Further adjustments after the mission has been started are rare in Tai chimpanzees, who tend to get it right first time. At Mahale, in Tanzania, this high degree of foresight seems to be lacking and the tool is gradually modified on an on-the-job basis as and when drawbacks in its efficiency become apparent. However, the ability to deduce a specific requirement and produce the tools to do the job has been impressively demonstrated by a female chimpanzee in The Gambia. This quick-thinking genius used nearby sticks to produce a succession of four types of tool to solve the problem of gaining access to the honey in a stingless bees' nest in a rather inaccessible site in a tree.

Chimpanzees also take much larger game than mere ants and termites. At Mount Assirak in Senegal the prey consists of some of their rather distant relatives, the Senegalese bushbaby and potto, both small nocturnal prosimians which sleep during the day when they are vulnerable to chance discovery and slaughter by foraging chimps. In other areas the quarry consists of diurnal monkeys such as red colobus; at Tai these are captured in a well-coordinated and apparently pre-planned drive. The sight of a chimpanzee gnawing away at the bloody, still warm remains of a red colobus after a hectic chase in which the unfortunate monkey was eventually cornered and torn apart shocked a television audience of millions when Sir David Attenborough's *Trials of Life* series was shown to an unsuspecting world in 1991.

At Gombe at least five species of primates succumb to the chimpanzees, with red colobus, red-tailed monkeys and blue monkeys being the most frequently hunted. Catching meat 'on the hoof', though, takes considerable cunning and a lot of effort, so the Gombe chimpanzees take a short cut when chance permits by letting another predator such as a baboon or leopard do the work first. The chimpanzees then appropriate the trophy from its rightful owner by ganging up and presenting a unified and intimidating front. Such mealtime mugging of leopards has been inferred rather than directly observed, but baboons are certainly robbed of their spoils as well as ending up on the chimpanzees' menu themselves. Most of the hunting is carried out by the adult males, who therefore end up claiming most of the meat; but they are not totally selfish and another member of the group willing to grovel sufficiently may receive some reward, as do any females who are in a sexual state exciting enough to make them obvious recipients of a meaty handout.

Most of the large savannah-living baboons also take vertebrate prey such as ground-nesting birds, hares, baby antelope and even other primates such as bushbabies and young vervet monkeys on an opportunistic basis. At least one group has acquired the knack of chimp-style cooperative hunting on a planned basis. This select band of Kenyan olive baboons outwits such fast and agile animals as small gazelles by driving them into the arms of their fellow baboons. As usual it is the males which take part in these hunts and a fruitful outcome may lead to some sharing of the spoils, a necessary adjunct to any cooperative venture in which all the participants must expect to share in the success of their joint enterprise. On the other hand kills made by solitary hunters were apparently not shared out before cooperative hunting began; after all, selfishness pays off when you have done all the work yourself.

At least two types of capuchins, the tufted and the white-faced, probably prey on other primates on an opportunistic basis: the white-faced's victims are the much smaller squirrel monkey, while the tufted capuchin, has once been seen preying on a dusky titi. Capuchins are the epitome of the successful generalist feeder, able to tackle just about anything within reason, both plant and animal. This includes such a surprising item as oysters on the seashore, which are successfully opened by white-fronted capuchins on Gorgona Island in Colombia. This makes an interesting parallel with the long-tailed macaque in Asia, also called the crab-eating macaque from its habitat of feeding on crabs and shellfish in mangrove swamps. African monkeys, too, do not miss out on the opportunity to add seafood to their diet, for riverine mangrove swamps in Senegal abound in small fiddler crabs which are a favourite food of the green monkeys (*Cercopithecus sabaeus*) which live there.

While searching for food capuchins really take their immediate neighbourhood apart in the most literal sense, ripping off any bark which might harbour something of interest such as a bird-eating spider, giant *Blaberus* cockroach or a nocturnal gecko in its daytime hideaway; breaking open hollow branches; picking over and examining large dead leaves; and delving around in the welter of debris which gathers at the bases of palm fronds. Part of the success of the white-faced capuchin in the tropical dry forests of Santa Rosa National Park in Costa Rica is due to the ability of neighbouring groups to make the best of whatever happens to be available in their particular neck of the woods. Thus one group spent over 80 per cent of its time feeding on fruit, an abundant resource in its area, and less than 20 per cent looking for insects which happen to be locally scarce. The stretch of forest next door was seemingly richer in insects, for the neighbouring capuchins divided their time almost equally between feeding on fruit and searching for insects. Further investigations seemed to indicate that these differences were directly related to the relative abundance of fruit and insects in the specific areas of forest, rather than to dietary preferences of the capuchins.

If a foraging capuchin happens to uncover a sleeping rat or mouse during its searching it will be eaten with relish. Such rodent-eating habits have also been seen in blue monkeys in Kakamega Forest in western Kenya. Most cercopithecine monkeys will routinely snap up any invertebrate prey they happen to come across – mostly small arboreal creatures such as geckos and other lizards, tree frogs, small snakes, birds' eggs and nestlings and even young squirrels. However, what makes the blue monkeys' hunting techniques at Kakamega rather unusual is their habit of descending to the ground in an explicit search for mice. Mouse hunting seems to be concentrated at certain times of the year, the monkey dropping into the undergrowth before emerging with its squirming victim. The monkey cuts short the mouse's struggles by bashing it against a branch, repeating this several times until close inspection reveals that all signs of life have evaporated. The blue monkeys of Kakamega are apparently connoisseurs of mouse meat, consuming only the best bits by fastidiously pulling out the guts and discarding them on the ground. Lizards, by contrast, do not demand such careful treatment and are merely stuffed down whole, tail first. A brief but seasonal shortage of fruit in the forest at certain times may account for the sudden concentration on meat eating, a phenomenon also noted in the closely related vervet monkey.

GUM

A substance produced by trees in response to injury, when gum first oozes forth it is soft and sticky; but on contact with the air it rapidly hardens, sealing off the wound. Primates take advantage of this facility in two ways: by inducing bleeding themselves using their teeth, which may be specially modified for the purpose; and by cashing in on secondhand supplies produced in response to damage by wood-boring insects such as longhorn beetle larvae.

As a diet gum has both advantages and drawbacks. It is relatively rich in carbohydrates and calcium and may even contain significant amounts of protein, plus reasonable quantities of trace minerals. It has the one great advantage over all other foods in being more or less completely predictable and reliable in its production. Even during the dry season a tree must produce gum to seal a wound – at this time other foods such as young palatable leaves, ripe fruits and insects may all be scarce or locally absent. The main drawback lies in the specialized abilities needed to digest gum, requiring modifications to the caecum. Primates which rely on gum for the greater part of their diet, such as *Cebuella, Callithrix, Phaner* and certain *Galago* species, all have the necessary gut modifications which enable them to utilize the nutrients in the gum to maximum benefit.

The fork-marked dwarf lemur is a perfect example of a gum specialist. The well-developed dental comb typical of lemurs has been extended forwards to act as a scraper. It works in consort with somewhat pointed upper teeth and an extra-long, flexible tongue to gouge out gobs of gum seeping from the galleries of wood-boring beetle larvae. Such secondary gum mines are not strictly necessary, as *Phaner*'s protruding dentition is the perfect tool for gouging holes in the bark to stimulate its own personal supply. Gum and sap constitute the bulk of the diet, with insects providing the only additional small-scale extras. Mouse lemurs also feed on gum but are unable to induce their own flows and rely on insect-created wounds.

The gum- and sap-feeding habits of bushbabies have been well studied in Africa. Most bushbabies eat gum as part of their diet, although the needle-clawed bushbabies such as *Galago elegantulus* seem to rely on it to a large extent. With their needle-like nails they are perfectly adapted for scurrying rapidly up and down vertical trunks and branches, busily making the rounds of the 500–1000 different gum stations scattered around their home range; both existing sites and new ones are sniffed out by their very sensitive noses. Only very small amounts of gum are licked up from each spot using the tooth scraper, which is apparently incapable of penetrating the bark to produce new wounds.

Although several kinds of tree capable of producing a reliable supply of gum are often present in a bushbaby's habitat, some are more attractive than others. In South Africa *Acacia karroo* seems to be the preferred gum source for *Galago moholi*, while in Kenya the gum of *A. xanthophloea* is preferred over that of *A. drepanolobium*. Vervet monkeys, which also often include gum in their extremely varied diet, also prefer *A. xanthophloea* over the *A. elatior* and *A. tortilis* which are available in the same kinds of habitat. The key seems to be the percentage of distasteful phenols present in the various gums; these are lowest in *A. xanthophloea* and highest in *A. tortilis*. Regular gum eaters are therefore expert gourmets, able to tell the good from the dubious at a sniff.

Marmosets exhibit several modifications for a gum-eating lifestyle. Having claws instead of the nails typical of other primates enables them to run with ease and confidence up and down the vertical trunks pockmarked with their gum wells. Their modified caecum copes well with digesting a gummy intake, while the rather short lower canines facilitate a gouging action on flat surfaces such as bark. By contrast, the projecting lower canines of the related tamarins *Saguinus* spp. would get in the way of such activities, restricting this branch of the family to lapping up exudates seeping from insect wounds on trees.

The pygmy marmoset (*Cebuella pygmaea*) shares the generally catholic diet of its group, taking a wide variety of insects, spiders, small fruits and tender young buds. However, although all the *Callithrix* species devote a great deal of time to collecting plant exudates, the pygmy marmoset devotes around 70 per cent of its foraging time to visiting the numerous holes which it has bitten in tree trunks around its territory. These lens-shaped lesions often pepper the bark of a large tree, for although the initial holes are made lower down the trunk, in time the marmoset gradually works upwards until the whole trunk appears to have been hit by a fusillade of grapeshot. The first two hours or so of each day are spent chiselling away at the day's fresh supply, while later in the day the marmoset ensures its long-term continuity of supply by carrying out the less immediately rewarding task of scraping out fresh holes.

FRUIT

Fruits can be divided into those which are designed to be eaten and their seeds disseminated by the consuming animal, and those which would prefer to be left alone to manage their own dispersal without being attacked and dismembered by some interfering bird or monkey. Primates eat vast quantities of both types, although the second entails overcoming certain difficulties such as a hard skin or chemical defences; to cope with these effectively may require various anatomical modifications. Even most animal-dispersed fruits employ some kind of dissuasive tactic to protect their unripe seeds until they are ready for action. This is often marked by chemical changes in the tissues surrounding the seeds, leading to a general softening of the fruit accompanied by a dramatic change to a more palatable flavour. At the same time there is often a striking adjustment in colour which calls attention to the newfound state of ripeness. Primates simply swallow small fruits whole without the luxury of preliminary chewing; larger fruits may demand a few quick bites. Although some seeds will end up being crushed by the animal's teeth, the majority probably survive their passage through mouth, stomach and intestines and are scattered widely around the consumer's home range in its droppings. Primates are probably important agents of seed dispersal for many tropical trees, and the seeds of at least some species are thought to germinate more readily after their chemical and physical scarification inside the animal.

The majority of primates include fruit in their diet, often varying the percentage of fruit versus leaves or fruit versus insects at different seasons. Thus a primate which appears to be a heavy fruit eater at one season may subsist almost entirely on leaves at another. Consequently in many cases it is rather dangerous in our present state of knowledge to make broad statements that a species is 'mainly frugivorous' or 'chiefly folivorous'. Feeding habits may also vary from one part of a forest to another, or from one habitat to another. This reflects the distinctly seasonal nature of the availability of ripe fruits in a particular primate group's home range, as well as the scattered and random distribution of fruiting trees at any one time within the area. In fact most monkeys are extremely good at knowing just when a particular tree's crop of fruits is going to be ready for consumption, and show an uncanny knack of turning up just at the right moment.

Such an example of perfect timing was evident when the author (KGP-M) was in Madagascar's eastern rainforests in April 1990. A large, compact clump of several hundred figs (*Ficus* sp.) was a prominent feature in a tree on the edge of the camp ground. On the first day these seemed untouched, but on the second day a black and white ruffed lemur showed up and began a beanfeast. This tree also lay within the home range of a group of white-fronted lemurs, who were most anxious to obtain their fair share of the now obviously newly ripe figs. But their efforts were regularly frustrated by the ruffed lemur, who showed every sign of claiming exclusive ownership of the fruits, barking loudly at any intrusion by the white-fronted lemurs and backing up its verbal warnings by lunging open-mouthed at any of their number who came too close. Although superior in numbers to the lone but bad-tempered guardian the smaller lemurs always gave way to the larger animal and retreated reluctantly into the forest.

The ruffed lemur engaged in long binges of fig eating, swallowing fruit after fruit in sessions lasting an hour or more. After gorging itself into a state of repletion it would then snooze off its excesses for a couple of hours on a branch directly above the figs, just handy for keeping a protective eye on them in case of a sneak raid by competing *albifrons*. Only when dusk fell did it retire into the forest to sleep.

Most primates enjoy eating figs. This black and white ruffed lemur (*Varecia variegata*) spent around a week defending its access to this clump of figs against an intruding group of white-fronted lemurs (*Eulemur fulvus albifrons*).

The persistent *albifrons* did manage to mount a rapid attack one afternoon when the piebald custodian was temporarily absent. Now, with their objective achieved, considerable intra-group rivalry arose over first access to the figs. The males, who had previously retreated without protest when threatened by the ruffed lemur, now lorded it over their females, whose attempts at getting anywhere near the figs while a male was in residence were greeted by an open-mouthed show of ferocity.

Under such an assault the entire batch of figs had gone within a week. Perhaps 90 per cent, maybe five hundred fruits, had ended up inside the single cat-sized ruffed lemur, whose capacity to cram down large numbers of these solidly built fruits at a single sitting was truly astonishing.

Now and again the ruffed lemur moved a short way to a large *Terminalia* tree which bore lots of almond-shaped green fruits. Dealing with these stubborn objects seemed to present major problems to the animal, which spent many minutes just trying to chew the hard outer flesh off the large inner stone, which, being too hard for the lemur to crack, was discarded. The amount of effort put into the chewing was so great that after one or two attempts the lemur would return to the far less intransigent figs, which quickly yielded to a few quick bites and a gulp. However, during figless periods the *Terminalia* fruits are probably worth the effort. Strangely enough, the resident aye-ayes with their huge powerful teeth can not only readily remove the rind of these fruits but actually do so in order to spit it out. They are after the stone inside, which they are capable of cracking and eating.

Figs are among the most popular fruits eaten by primates in the tropics, and a ripe bunch rapidly draws a crowd of admirers eager to make quick work of the delicacies and distribute their myriad tiny seeds throughout the forest. Ripe figs can be so attractive that during the fruiting season in the Ankarana massif in north-west Madagascar groups of crowned lemurs carefully make their way across the spectacular wilderness of foot-shredding, splinter-like limestone pinnacles known locally as 'tsingy' to reach the fruiting fig trees which grow in this hostile landscape.

Fruits of various kinds represent a key resource for this endangered lemur at the end of the long dry season, when many of the trees are leafless, although some of the available fruits require careful handling if disaster is to be avoided. For example, the green fruits of a common *Strychnos* liana contain highly toxic seed which must not be ingested by the lemur. The solution is to strip away the outer peel using the teeth, and then carefully to chew the desirable flesh from around the poisonous seed, which is simply dropped. Like most primates the crowned lemur is very fussy about which fruits are acceptable and which are rejects.

Primates have a keen sense of discrimination and will rapidly sniff their way through a branch full of fruits, eating some and rejecting others in rapid succession, even though to our eyes the whole lot may look identical. This could be partly because many fruits are acceptable while still far from ripe, for any animal which wants to get the maximum profit from what is only temporarily on offer is faced with a conundrum: either wait until the fruit is

Lemurs seem to grade the quality of fruit by a rapid inspection with the nose. This female white-fronted lemur (*Eulemur fulvus albifrons*) rapidly sniffed her way along this *Grewia* branch, pulling each bunch of fruits towards her face and eating some while rejecting others.

properly ripe and risk losing it to some competitor with no scruples who got there first – and the competition includes many birds and bats as well as other primates; or else beat the opposition to the draw by downing as much as possible while it is still unripe. This does, however, mean taking the risk of suffering indigestion from any defensive chemicals which may be present in the fruit in its unripe state.

Most primates seem to attempt a compromise by strictly grading the fruit by nose and eating those which are just about ripe enough to be edible. This testing procedure can be a wasteful and messy business, as many monkeys pick the fruit first rather than smelling it while still on the branch, and then use a quick bite rather than a mere sniff to assess its status. Rejects are simply dropped and the continuous rain of debris, up to grapefruit size, which plummets earthwards can make primate watching below a somewhat hazardous occupation (even if the animals themselves don't decide to pelt you with sticks and urinate on your head, as sometimes happens).

Yet such apparently profligate feeding habits may actually be regarded as a welcome form of free handouts by earthbound consumers, who would otherwise have to wait around for the fruit to fall by itself. In India the Hanuman langur is known locally as the chital's herdsman because a group of these beautiful spotted deer (*Axis axis*) is such a common sight beneath a tree in which a group of langurs is noisily foraging. Chital and also, on occasion, wild boar (*Sus scrofa*) hang around, cleaning up all the half-bitten fruits so conveniently dropped as a result of the langur's sloppy feeding methods. As neither deer nor pigs can climb trees this association is obviously of great benefit to them, especially as by the time the fruit would have ripened and fallen naturally the langurs and other arboreal guzzlers would have cleaned most of it out anyway. Such an association also benefits both sides by providing a mutual defensive system – deer and langurs respond to each other's alarm calls when the sharp eyes of the langurs or keen noses of the deer detect the stealthy approach of a tiger or leopard.

In South and Central America the capuchin monkeys seem to be the most skilled performers in the art of knowing when fruiting and flowering trees are at their best. They regularly but unintentionally lead other less proficient species such as squirrel monkeys to the source of plenty, although the capuchins take priority in feeding when such mixed-species groups occur. Where brown and white-fronted capuchins occur together, it is the latter which tag along behind the pathfinding browns, only to throw their weight around and appropriate the best feeding spots when they arrive.

The mountain gorilla's ingestion of vast amounts of the vegetative parts of plants such as leaves and stems (around 90 per cent of their total intake) in the cold, wet, mountainous habitat of the Virungas volcanoes had given the impression that gorillas in general are strongly folivorous. However, some recent studies of a population of the western gorilla in warm lowland disturbed forest in Gabon point to their taking a highly frugivorous diet, for more than 97 per cent of their droppings contained the remains of fruits belonging to at least seventy-two different species. Of the larger apes, orang-utans too seem to rely heavily on fruit, although at certain seasons they switch to other forest products as part of their incredibly varied diet. This may include as many as 317 different food types, from fruit to fungi, insects and honey.

SEEDS

The seed or seeds contained within a fruit, rather than any pulpy flesh which may surround them, are the target for a select band of primates which boast the necessary hardware to gain access, which can be difficult. Sometimes, however, this is not much of a challenge and the barbary macaques, which rely heavily on acorns to keep them going during the autumn, have little difficulty in chewing them up. But many fruits protect their vital seeds within an incredibly hard outer case designed to be impervious to attacks by frugivorous animals. The resistance of this outer layer to assault by straining jaws is the governing factor determining which primates can eat the seed within and which cannot. For example, bearded sakis are able to pierce the outer protective pericarp of fruits with a puncture resistance fifteen times as great as those which can be opened by black spider monkeys in the same forest. The key to success lies in the bearded saki's can-opening dental arrangement, with the upper set of front teeth jutting forwards to drop over the lower set like a macaw's beak – an apt comparison, for macaws are the champion nutcrackers of the avian world. Bearded sakis manage to open fruits similar to Brazil nuts; a human would need a hammer to do likewise. Uakaris too are specialized seed-eaters, and have similar dentition and capabilities. Both unripe fruits and those kinds which are not designed to be eaten by animals often protect their seeds inside a poisonous fleshy outer layer which is carefully stripped away by the uakari before the naked seed is mangled between the formidable teeth. In Africa the black colobus devotes over half its time to feeding on seeds, while in Asia the proboscis monkey is a specialized consumer of the seeds of a number of the dominant trees in its habitat.

Seed swallowing does not always have the desired outcome, however. In Madagascar ring-tailed lemurs regularly eat *Nestina* seeds, only to vomit them up again a few minutes later in a violently emetic reaction which is apparently provoked by a defensive chemical coating on the seeds. This presumably does not adversely affect their flavour, but makes it impossible for the ring-tail to keep them down. Yet the Verreaux's sifakas which live in the same forests regularly eat these seeds with no ill effects. Sifakas have also been seen eating the leaves of such notoriously poisonous plants as members of the Solanaceae (tomato and potato family), also with no apparent adverse reactions. Such variations in the ability to cope with the chemical defences of plants are often seen between different primate species sharing the same habitat, and probably help to reduce competition between them for available resources.

Monkeys which lack nutcracking teeth may use their intelligence to try to solve the problem. Green monkeys may spend considerable periods sitting on the ground, a hard-shelled fruit in each hand, repeatedly bashing them together in the hope of breaking in. This is such an established occupation that the gentle tap-tap-tap is a common accompaniment to a walk in forests where this monkey occurs. Unfortunately the success rate seems to be low and the hopeful tapper usually throws the nuts down still stubbornly intact, only to pick up another two and try again. Tufted capuchins in South America are known to employ similar techniques with considerable success, varying their methods by also smashing the nuts hard against a solid tree trunk.

The prize for non-dental nutcracking expertise must however go to the chimpanzees at Mount Tai; they are far more successful in their endeavours, applying their superior intelligence to the task to come up with the correct tool for doing the job properly. This consists of a rock weighing as much as 20 kg (44 lb) – nothing much smaller is of any great use in opening the large, super-hard panda nuts. An alternative but less favoured tool is a piece of wood which, if not quite the right size for the job, will be fine tuned until it is. The chimpanzee does this either by bashing it against a convenient protruding root or by levering upwards until it breaks at the right spot. Suitable stones are scarce at Tai and have to be carried, usually along with a supply of nuts, to an 'anvil' which tends to be a protruding tree root. During the main coula nut season Tai chimpanzees indulge in nutcracking sessions for an average of two

Grivet monkeys (*Cercopithecus sabaeus*) in the Abuko nature reserve in The Gambia spend much of their time foraging for fallen fruits on the forest floor. They sit for ages with a nut in each hand, tapping them together in an effort to open them. When this fails they resort to the use of their teeth, although with the harder nuts neither method may yield results.

and a half hours each day, although certain devotees may carry on thumping away for five hours or more. At this season coula nuts provide the bulk of the daily intake of calories, making them a major resource; yet for some unexplained reason it is ignored by other groups of well-studied chimps such as those at Gombe or Mahale, which have never been seen cracking open hard nuts using hammers.

The ingenuity of the Tai chimps does not stop at using a tool solely to open the nut; a second tool, a small stick, may then be employed to winkle out the deeply embedded kernel of nuts such as *Detarium senegalense* which is inaccessible to the chimpanzees' best efforts at gnawing. Similar methods may also be used to winkle out the brain from the skull of a colobus monkey. Tool using of any kind requires many years of trial and error before perfection is achieved, so young chimpanzees start early by studiously watching their parents at work, acquiring the theoretical knowhow before they themselves make their first fumbling attempts at the actual manual side of things.

The aye-aye is aberrant in so many ways that it is not surprising that it has evolved its own special method of gaining entry to intransigent nuts such as the ramy, whose three-chambered interior means that a single bite to crack the shell and extract the contents simply isn't good enough. So the aye-aye gnaws away a hole with its chisel-like teeth and then utilizes its long skeletal third digit to scoop out the contents, very much as chimpanzees use a stick. The technique seems to be so successful that piles of these nuts, each bearing the telltale hallmark of an aye-aye's breaking and entering methods, are often found beneath ramy trees.

FLOWERS

Most primates include at least a few flowers in their diet at various times of the year, but when a brief abundance of flowers coincides with a shortage of tender young leaves, fruits or insects, they may temporarily rise to the top of the menu. Few primates visit flowers in the beneficial way seen in bats, which are important (often the sole) pollinators of many nocturnal flowers. By contrast, primates tend to have entirely negative relationships with flowers, partly or wholly devouring the reproductive structures and thus preventing fruit developing.

There are a few exceptions, mainly involving small species which visit flowers mainly or solely to lick up the nectar and pollen. In so doing pollen may be picked up on the face and transferred to another flower on the same or an adjacent tree. In Madagascar the red-bellied and red-fronted lemurs both visit the large 8 cm (3 in) long fleshy flowers of a *Strongylodon* sp. tree and probably act as effective non-destructive pollinators. The red-bellied lemur also visits some rather less robustly built flowers in the family Sterculiaceae and in licking up nectar and pollen probably assists in pollination, yet the same flowers are torn to shreds and eaten by the rather larger red-fronted lemurs. At least eighteen species of flowers are visited by these two lemurs, all but the above-mentioned examples resulting in the destruction of the flower. During January and February flower feeding takes up 40 per cent of the feeding time of these two lemurs, coinciding with a lull in the production of ripe fruits (the preferred food throughout the year) and young palatable leaves. During July the red-bellied may feed extensively on *Eucalyptus* flowers (80 per cent of its diet at that time), alien trees usually present as tall emergents in the patch of forest where the study was made. The rest of the July diet comprised 10 per cent fruit and 10 per cent young leaves, while just a month earlier fruit had been the major food; this illustrates the propensity shown by many primates for switching from one resource to another depending on availability.

Interestingly enough, *Eucalyptus* flowers were invariably consumed at night,

although the red-bellied lemur had always been considered to be diurnal. However, it has recently been recognized that a number of lemurs have a twenty-four-hour activity cycle in which bouts of feeding and resting are alternated throughout both day and night. Such a routine has been entitled 'cathemeral' and has so far been observed in the red-bellied, red-fronted, mongoose and crowned lemurs. The mongoose lemur divides up the percentage of its activities allocated to either day or night according to season, tending to be mainly nocturnal during the flowering period of the kapok trees (*Ceiba pentandra*) whose sweetly scented nocturnal white flowers form the bulk of the diet for the duration of their appearance. Like most bat-pollinated flowers those of the kapok are rich in nectar, and the mongoose lemurs dart rapidly around the branches delving their long noses into the flowers to lap up the sweet supply, becoming covered in pollen in the process. The flowers themselves are not eaten, probably because they are poisonous. The black and white ruffed lemur has also been seen acting as a possible pollinator of rainforest flowers, and in South America tufted capuchins, saddleback tamarins and a few other species have been seen regularly licking nectar from *Combretum* flowers.

LEAVES

These form at least part of the diet for many primates, but only a few are specialized for eating little else. At first sight leaves would appear to be the easiest of foods to obtain. After all, most primate species live in tropical rainforests where a mass of foliage is available throughout the year, providing a seemingly inexhaustible and ubiquitous source of nourishment to be had merely for the picking. Life, however, is seldom that simple, and foolish indeed is the primate who just plucks the nearest available leaf. Any old leaf simply will not do, for leaves are complex entities and the undiscriminating consumer could have to put up with anything from an acrid taste to intense vomiting or even severe illness and death.

Such possible traumas come about because trees are not happy about having animals eating their leaves. Survival can be tough enough for a tree anyway, and the extra burden of freeloading animals battening on to precious leaves is something to be avoided at all costs. As it is impossible for a tree to uproot itself and run away from such attacks, many kinds have adopted a suitably static defence – chemical warfare. Even innocuous-looking leaves are often laced with defensive compounds which make them too unpalatable to swallow or unhelpfully unprofitable to eat. The two main compounds are alkaloids, which may induce severe poisoning, and tannins which act as digestion inhibitors. The latter probably function by making the environment in the gut hostile to the symbiotic bacteria which are so vital in digesting the cellulose content of the leaves. As this, when broken down, contributes most of the leaf's nutrient content, such inhibitors are a potent method of defence.

Some primates are more resistant to these chemical counter-measures than others. The ability of sifakas to cope with items which poison ring-tailed

Leaves form the bulk of the diet for some monkeys, such as these silvered leaf monkeys (*Presbytis cristata*) in a Malaysian forest.

lemurs has already been mentioned. Red-fronted lemurs are better able to cope with secondary defensive compounds than red-bellied lemurs, reducing competition in the rainforests where the two species live close together. Hanuman langurs are renowned for being able to feed on leaves which are avoided by other animals, even by insects which are the supreme exponents of the art of mastering such defences. But the capacity to avoid being poisoned by one's food reaches its zenith in the golden bamboo lemur.

This rare animal feeds almost solely on the leaf bases and fresh growing shoots of the giant bamboo (*Cephalostacyum viguieri*). These parts of the plant, and especially the pith of the young shoots, contain high concentrations of hydrogen cyanide – in a day's normal browsing a golden bamboo lemur will

This female golden bamboo lemur is tucking into a young stem of the giant bamboo, regardless of the high cyanide content. Just how this animal avoids being poisoned is currently unknown.

take in around twelve times the normal lethal dose for an animal of its size and weight. Quite how it accomplishes the amazing feat of avoiding the symptoms of acute cyanide poisoning is currently unknown; the explanation will no doubt be most interesting. The greater bamboo lemur, another very rare animal, occupies the same forest and eats the same species of bamboo but avoids its golden relative's apparent dicing with death by only selecting the mature culms which are free of cyanide. The grey bamboo lemur is also present, but mainly feeds on a different, less hazardous bamboo (*C. perrieri*) which does not contain cyanide at all.

Few primates bother with tough old leaves, as these are irksome to chew and difficult to digest. There are exceptions, such as the black and white colobus which during hard times can happily cope with a high proportion of old leaves in its diet. Generally speaking, though, young tender leaves are preferred, and as these are widely scattered in the forest and seasonal in appearance searching them out uses up a lot of time. The largest Madagascan lemur, the indri, only eats young leaves, carefully picking over a branch to locate those fresh tender examples which most suit its fussy tastes. The West African olive colobus is similarly fastidious, virtually ignoring mature leaves in favour of immature specimens plus a few fruits, flowers and seeds. However, in extreme habitats, where living conditions are on the borderline and choice of foodstuffs limited, plants which would otherwise be ignored may of necessity form the staple diet. Thus the tufted capuchin is a widespread and highly adaptable species which usually gleans a varied diet containing plenty of very nourishing high-value foods such as fruit and insects. However in the El Rey National Park in Argentina, a marginal habitat for this monkey, its diet consists mainly of rather tough-leaved epiphytic bromeliads. These are nutritionally poor but have the advantage of being so common that they are there for the taking at all times of year, including the dry season when there is precious little else. So this highly versatile monkey makes do with chomping away on a monotonous progression of bromeliads and manages to make a living in a basically hostile environment.

A few primates seldom if ever feed on the leaves of trees but instead spend the day picking herbs growing at ground level. The rhesus macaque seems to prefer grasses and clovers over all other foods. Such plants typically cover open areas where the original forest has been felled or disturbed, which may account for the success of this monkey in recent years while humans have been altering the environment in its favour. Yellow baboons and, in some parts of their range, olive baboons also live on a grassy diet. However, the main exponent is the gelada, which has developed grass plucking to a fine art. Geladas inhabit the cool, high mountain plateaux in Ethiopia, a land of few trees but abundant grasslands. Here the gelada spends the day shuffling along on its bottom, laboriously yet skilfully reaping by hand the corms, blades or seeds (the proportion of each depends on seasonal availability) of the ubiquitous grasses and stuffing them into its mouth.

Leaves as medicines?

A strange example of exploiting the defensive chemicals in leaves in a positive and beneficial way has been observed in chimpanzees. Several times these intelligent animals have been seen holding the leaves of an *Aspilia* species between their tongue and the mucous lining of the mouth. This would be an efficient way of absorbing medicinal compounds through bypassing enzymes in the gut or liver which might destroy them before they could enter the bloodstream. *Aspilia* leaves are used by the local people in native medicine and contain high concentrations of the chemical thiarubrine which acts powerfully against bacteria, fungi and helminths. A rather ill-looking chimpanzee has also been observed sucking the bitter sap from the pith of *Vernonia amygdalina*. With its unpleasant acerbic flavour this plant is seldom eaten by chimps, but is often exploited by the local people to cure stomach and intestinal disorders.

SOIL AND WATER

All primates drink when water is available, even if that only means licking the early morning dew off their own fur, a favourite habit of Verreaux's sifakas who are otherwise not thought to visit standing water. This is probably because sifakas, along with a few others such as colobus monkeys, can metab-

Savannah baboons drink regularly. This chacma baboon (*Papio ursinus*) stoops to drink at a waterhole in the enormous Kruger National Park, South Africa.

Several species of lemurs regularly visit certain spots to eat the soil itself. This is a female red-fronted lemur (*Eulemur fulvus rufus*) at Ranomafana. The exact purpose of this strange custom is not known.

olize sufficient water from their leafy food to avoid drinking for long periods.

Many primates simply pay a regular visit to a lake or river and dip their mouths directly into the water to drink. However, because of their skill in manipulating objects in their hands primates are able to make use of small, naturally occurring containers such as scoop-shaped dead leaves. These are picked up, the head tipped back and the contents poured into the mouth.

Lemurs, monkeys and apes are also adept at finding small reservoirs of water trapped inside tree holes, a useful fallback during the long dry season. If the entrance hole is large enough to admit the whole head, then the contents are slurped up directly in the conventional manner; but if the hole is too small or the water level falls too low to be reached by a craning neck then both monkeys and lemurs will repeatedly dip a hand inside and lick off the resulting wetness. Chimpanzees have gone one step further, increasing the amount of water which can be extracted at one dip by shoving a handful of leaves inside to act as a sponge.

Crowned lemurs in Madagascar's Ankarana massif go in for rot hole hand dipping, but also take advantage of a permanent but rather unusual water source during the dry season by climbing 5 m (16 ft) down a gloomy vertical shaft in the rock to drink at a subterranean pool. The habit has been followed for so long that the rocks leading down to the water are polished by the passage of generations of thirsty lemurs. Sanford's lemurs share the same dry forests, yet have never been seen to venture into the depths to slake their thirst.

Several kinds of lemurs, including the red-bellied, red-fronted, ruffed, Verreaux's sifaka, diademed sifaka and indri, regularly come down to the ground in certain well-used spots to lick up the bare earth, sometimes on an almost daily basis. Why they do so is not known; it may add something to a minerally deficient diet or may help counteract digestion-inhibiting leaf toxins.

Chapter 7

Habitat and Ecology

Unlike such supremely successful animals as the rodents, the primates are not found virtually everywhere. In fact a glance at a primate distribution map makes it immediately obvious that there is an enormous concentration within the tropical regions, although even here certain specialized or highly adaptable species are found far up into the mountain within a temperate climatic regime. Vast areas of the temperate world are unfortunately devoid of primates, including the whole of Europe, the vast swathes of Siberian coniferous forests, the whole of North America north of southern Mexico's tropical regions, the extensive beech forests of southern Argentina and Chile, and Argentina's vast expanse of grassy pampas.

RAINFOREST AND ASSOCIATED HABITATS

Even within the tropics a single environment, rainforest, provides the main or sole habitat for the bulk of the primate species. This should not perhaps come as a surprise, given that these same forests are home to the overwhelming majority of the planet's total wealth of fauna and flora. In fact a few hectares of Amazonian rainforest can boast more species of trees or butterflies than the whole of Europe. Rainforest presents a very stable environment to its animal denizens, with year-round temperatures which vary little between the seasons, which themselves may be marked more by differences in rainfall than by any fluctuations in temperature. In fact day and night temperatures vary more than annual ones, so that in many of these forests the leaves are able to remain on the trees throughout the year. However, even in the tropics some rainforest areas are semi-deciduous, and many of the trees will drop their leaves during a noticeably drier period. Even this may not put too much pressure on primate populations, as some trees may ripen their fruits during the dry season. Insects may also be more abundant and easily found at this time than during the torrential height of the rainy season. For example, in Corcovado National Park in Cost Rica invertebrate numbers were found to reach their lowest point towards the end of the rains when the forest is at its lushest.

Nevertheless, with such an all-encompassing variety of potential foodstuffs at hand and faced by none of the severe food shortages typical of the long temperate winter, it is not surprising that a dozen or more species of primates can coexist in a single area of rainforest in both the Old and New Worlds. To illustrate this it is worth looking at a few of the better-known primate 'hot spots' to see how specific areas within the different continents figure in the world primate habitat rankings.

The area of the M'Passa Plateau near Makokou in northern Gabon is often held to be the single richest locality for primates in the world, with fifteen species recorded (although two, the gorilla and grey-cheeked mangabey, have recently been rendered locally extinct). Five of these are nocturnal prosimians such as bushbabies and the potto; there are four different *Cercopithecus* spp, plus the talapoin, chimpanzee and mandrill. The lack of even a single species of colobus is surprising, given the presence of at least one species of these easy-to-please leaf eaters over vast areas of Africa. Yet just 40 km (25 miles) to the north-east of Makokou lies a swamp forest alongside the Liboy River where the black and white colobus is present. The tropical rainforests which clothe the Tai National Park in the Ivory Coast do, however, seem to provide a suitable home for these leaf eaters, as three species are present – part of the complement of eleven different primates known to occur here. Among the fourteen species found in the beautiful Korup reserve in western Cameroun are the forest drill and the very rare L'Hoest's guenon (*Cercopithecus lhoesti*). In this amazing forest it is possible to sit quietly and watch as many as seven different species of diurnal monkeys going about their business at the same time in the trees above.

Turning now to the nearby mini-continent of Madagascar, areas of the eastern rainforests here easily rank among the world's top primate habitats. The Strict Nature Reserve of Adohahela boasts no fewer than twelve species of lemurs in its 760 sq km (nearly 300 sq miles). However, not all of these are confined to rainforest, as this reserve uniquely straddles a vegetational divide. Species such as the fork-marked lemur and grey mouse lemur inhabit the dry 'spiny forest' segment of the reserve, which lies adjacent to a tract of rich rainforest containing the rest of the species. These include the rufous mouse lemur, making Andohahela one of the only places where both the closely related species of mouse lemurs live cheek by jowl.

The inclusion of two forest types within a single area does not, on the other hand, account for the richness of the forests around Ranomafana on the edge of the eastern escarpment. Here large areas of rainforest densely envelop rolling hills sloping down to a foaming river, although much of the more accessible forest is secondary in origin. Yet in both primary and secondary formations there are lemur species in plenty, with no fewer than twelve different kinds sharing the same habitat, including all three species of *Hapalemur* bamboo lemurs. Two of these, the golden and greater bamboo lemurs, are known for certain only from the forests around Ranomafana and are numbered among the world's rarest and most endangered primates. Their restriction to this small area is puzzling, as their main bamboo food plant is more widespread and similar-looking forest still occurs over quite a large area.

Further north, the rainforest in the well-known reserve at Perinet hosts nine lemur species in a mere 810 ha (2000 acres), including the largest and most spectacular of all, the indri. Just outside the reserve boundary occur the black and white ruffed lemur and the spectacularly beautiful diademed sifaka, both of which, it is hoped, will soon recolonize the reserve itself, raising the total to a very respectable eleven species.

The tropical rainforest near Ranomafana in south-eastern Madagascar is home to twelve species of lemurs including the very rare golden and greater bamboo lemurs (*Hapalemur aureus* and *H. simus*).

Further east, the Asian dipterocarp rainforests mainly occur in peninsular Malaysia and on the islands of Borneo, Sumatra and Sulawesi. The last-named, relatively small island is noted for containing seven macaque species which are found nowhere else. The forests of Kuala Lompat in west Malaysia also hold seven (non-endemic) species, while in Indonesian Kalimantan on the great island of Borneo the reserve at Kutai can boast eight. Yet continue on eastwards into the rainforests of New Guinea and Australia and you will find that primates are absent, both these large landmasses having split away and drifted out of reach long before the first primates had come on to the scene. Tree kangaroos now occupy the primate niche to a certain extent in the rainforests of New Guinea and tropical Queensland.

The most productive general area for primates is probably the region of Upper Amazonia, where up to twenty-two species may be found in a zone of rainforest covering 500 sq km (190 sq miles). Perhaps the richest as well as the best-known locality here is the Cocha Cashu sector of Manu National Park in Peru, where thirteen different primate species forage in the hot, steamy lowland rainforest around the Cocha Cashu Lake. Many of these feed on the same type of fruits, leaves and insects, and it is a matter of intensive study and some speculation as to how wasteful competition between species is avoided when such large numbers of primates coexist and compete for the available resources. In certain cases different species forage at different levels in the forest, or utilize different foods at different times of the year, or search for them in a different way which tends to target different species.

Having said that, there are great similarities in the diets of large numbers of

Coastal mangroves in Borneo are the favoured habitat for the bizarre proboscis monkey (*Nasalis larvatus*).

primates as diverse as monkeys in Manu and lemurs in Perinet, while the competition between primates and the plethora of fruit- and insect-eating birds and bats complicates the picture still further. Even the leaf eaters have to contend with competitors trying to get there first, such as the occurrence, sometimes in high densities, of leaf-guzzling sloths (*Choloepus*, *Bradypus*) in primate habitats in South and Central America, and of similarly folivorous tree hyraxes (*Dendrohyrax*) in some African forests.

Whatever the level of competition, however, one thing is certain: primates are usually very successful forest dwellers. They are major predators on the leaves and fruits of the forest trees, as well as taking a heavy toll of the invertebrate and, to a lesser extent, vertebrate life around them. The actual weight of living monkey meat swinging around in the forest may in fact reach quite remarkable levels – more than 3500 kg per sq km has been recorded in Uganda's Kibale Forest, made up of ten species of primates. This weight of flesh supported by the forest ecosystem is nearly 90 per cent of the total weight of the antelope and other large ungulates sustained by the Serengeti ecosystem in Tanzania, a region famed for the awesome density of its grazing herds. The weight of primates per unit area is lower in other rainforest but still reaches remarkably high levels. This should not perhaps be unexpected, given the unique ability of primates to utilize a wide variety of foods at every level of the forest, from the ground floor to the topmost branches of the tallest emergent trees up to 90 m (280 ft) high.

In view of their great bulk it is hardly surprising that mountain gorillas spend most of their time on the ground. The preferred habitat is regenerating secondary forest or montane forest; the latter may consist of dwarf elfin forest

Geoffroy's spider monkey (*Ateles geoffroyi*) has to compete with leaf-eating sloths in its rainforest habitat in Panama.

permanently shrouded in mist and more or less perpetually cold, wet and miserable. Dense forest is less favoured than a more open type in which plenty of light can penetrate the canopy; this results in a luxuriant growth of herbaceous vegetation which provides the gorillas with their main source of food. The constant feeding pressure exerted by the placid heavyweights as they seek to satisfy their giant-sized dietary needs by chomping away for long periods on such favourite food plants as the giant parsley (*Peucaedanum linderi*) and the thistle *Carduus nyassanus* may actually promote the ongoing production of tasty new shoots. So their eating habits may in the long term actually increase the food supply rather than decrease it. The rate of appearance of new shoots and the rate of growth of various *Afromomum* spp. gingers also seems to be encouraged by the gorillas' inroads. Although the newer shoots are higher in

Mountain gorillas (*Gorilla gorilla beringei*) munch their way through huge quantities of green fodder each day, stimulating vigorous regrowth in some of their preferred food plants and helping to maintain the rather open nature of their favoured habitat.

nutritional content, the mountain gorilla also has to pack away large quantities of less nutritional fibrous food such as tough old stems and bark, spending up to half the day, doggedly chewing away on its vast daily intake. After such mammoth feeding sessions it is perhaps only to be expected that the gorillas devote most of the rest of the day to dozing it off; after all, digesting such large amounts of fibrous food is not conducive to much energetic activity.

Lowland gorillas enjoy far warmer conditions and a quite different diet, eating far more fruits than their mountain counterparts. High populations of eastern lowland gorillas have recently been found in the rather unusual habitat of the Likouala swamp in the north-eastern Congo. This is one of the largest swamps in the world, covering around 63,500 sq km (24,500 sq miles) at a comfortably warm elevation of 300–350 m (1000–1200 ft). The forest is seasonally flooded up to a depth of 1 m (3 ft) and the underlying soil remains permanently saturated throughout the year.

Similar but far more drastically changing conditions also exist in the flooded forests or varzea on the upper Amazon, covering an area approaching the size of the British Isles. Here the seasonal flooding may last six months or more, during which the waters rise as much as 10 m (30 ft) up the trunks of the trees, often leaving only the crowns exposed and completely submerging seedlings and saplings. This periodically inundated forest is the home of the white uakari monkey which is found nowhere else, not even in the luxuriant non-flooded rainforest surrounding the varzea. Why this should be is not known, but it may be connected with the uakaris' habit of eating tough-skinned fruits. These are found on the trees during the wet season and as sprouting seedlings on the ground after the floods have receded, when the uakaris come down to ground level to harvest the fresh crop of germinating saplings.

Perhaps the most unusual watery habitat for any primate comprises not swampy forest but the open, treeless reedbeds around Lake Alaotra in Madagascar, the sole habitat for the rare and endangered ssp. *alaotrensis* of the grey bamboo lemur *Hapalemur griseus*. This lemur usually inhabits rainforest, feeding at all levels from the high canopy to the forest floor on bamboos, which are their main food; secondary forest in which a vigorous growth of bamboos often predominates may support even greater densities. However, the open lakeside reedbeds at Alaotra are bereft of bamboos; in fact they are such a far cry from such forested habitats that it is surprising that a population of this lemur should ever have become established there. Reeds and sedges substitute for the absent bamboos as the animals' main food. Unfortunately for the lemurs the reedbeds dry out towards the end of the dry season when they are easily set on fire by the local people, driving the terrified lemurs out of their home into the waiting arms of the villagers, to be clubbed to death and eaten.

TROPICAL DECIDUOUS FOREST

After rainforest the habitat type most able to support the highest density of primates, some of which may be restricted to this formation, is variously known as 'seasonal forest', 'tropical dry forest' or 'tropical deciduous forest'. Whatever the name used, this type of forest is characterized by having generally shorter trees with smaller leaves, which are dropped during a dry season which may last for six months or more. Large areas of this type of forest occur in western Madagascar, South and Central America, East Africa and the monsoon forest which covers extensive areas of India and Sri Lanka. Primate diversity is lower than in rainforests due to the constraints on survival imposed by the more seasonal nature of the food supply. Even so, primate numbers may be quite respectable – for example, no fewer than seven species of lemurs thrive in the Ankarafantsika Forest reserve in north-west Madagascar. During the long, hot dry season shortage of food in the less favourable areas of these forests may force most of the resident primates into unusually close contact within the more productive sections.

By the end of the winter drought in the Ankarana massif in north-west Madagascar many of the resident crowned and Sanford's lemurs are congregated in the Canyon Grand and Canyon Foristier. They are attracted to these verdant oases of lusher growth, with many tall trees and a rampant growth of vines, all kept more or less green throughout the year by an underground stream. Here the crowned lemur may reach a density of around 500 animals per sq km (1300 animals per sq mile). This is a remarkably high figure for such

The beautiful Coquerel's sifaka (*Propithecus verreauxi coquereli*) is restricted to the tropical dry forests of north-west Madagascar such as the forest reserve of Ampijeroa.

a seasonal habitat, although nowhere near the highest recorded figures of 1200 per sq km (3120 per sq mile) for the red-fronted lemur and 1000 per sq km (2600 per sq mile) for the ring-tailed lemur. Both these figures were recorded in dense, high-canopied south-western gallery forest dominated by large tamarind trees, whose lavish supply of much-relished fruits probably attracts such high densities of lemurs at the impoverished end of the dry season like bees to a honeypot. Weasel or sportive lemurs are also abundant in the Canyon Grand at Ankarana. Shortly after nightfall the males can be heard squawking and chattering loudly as they call the odds at one another from adjacent territories

The rare mongoose lemur (*Eulemur mongoz*) is found only in scattered areas of tropical dry forest in north-west Madagascar and on the Comoro Islands.

at intervals of around 30 m (30 yds) or so. This represents a very high density for this nocturnal folivore, the most vocal of the nine species of lemurs found in this spectacular habitat.

By comparison, the monsoon forests of the Indian subcontinent usually only manage to muster around three to four species, and the latter figure is the number found in the Polonnaruwa sanctuary in Sri Lanka. Despite its seasonal nature the environment here is able to support the highest density of primates so far recorded in a dry forest, at 2840 kg of monkeys per sq km (7380 per sq mile). The availability of handouts from humans and the proximity of nearby farms which can be raided may explain this exceptionally high figure.

No primates are restricted to 'woodland' and 'thornwood', which generally occurs in areas with very seasonal or even sporadic and unreliable rainfall. Many of the plants defend themselves with spines, the acacias being typical examples, and there may be such large gaps between the trees that there is no continuous canopy enabling a primate to move long distances without coming down to the ground. This kind of woodland occurs widely in both East and West Africa, over a large area to the south of the Amazonian forests in South America, and in parts of south-east Asia. However, the most intriguing and indeed unique example of this formation is the so-called 'spiny forest' which still covers large areas of southern Madagascar.

The most interesting feature of the spiny forest is the fact that the plants

The tropical dry forests around the spectacular limestone massif of Ankarana in north-west Madagascar host large populations of the rare crowned and Sandford's lemurs (*Eulemur coronatus* and *E. fulvus sandfordi*).

which generally dominate the landscape in their thousands and play a major role in the economy of the animals living there belong to a family called the Didiereaceae which, being one of this island's several endemic plant families, is restricted to this region of Madagascar. A number of the most dominant species of *Alluaudia* and *Didierea* resemble tall candelabra cacti, lending the whole landscape a resemblance to the American West or Mexico. Unlike their cactus look-alikes, however, these Madagascan plants bear numerous small deciduous leaves which clothe their stems during the wet season, when they provide the main source of food for the resident weasel lemurs.

Although to outward appearance the spiny forest seems a hostile and unproductive environment for any primate, a single area may in fact provide

A male Sanford's lemur (*Eulemur fulvus sanfordi*) distinguished by his pale beard and side whiskers.

satisfactory board and lodging for as many as five different lemur species. This is a high number for such a dry, semi-desert habitat, and one which similar areas in other continents probably cannot approach. However, the actual animals are thin on the ground compared with the numbers found in rain-forests, deciduous forests or even the rich tall growth of gallery forests, which skirt the rivers meandering through the heart of the spiny forest itself. One of the great mysteries of the spiny forest is how the resident Verreaux's sifakas and ring-tailed lemurs manage to leap quite confidently from one heavily spined trunk to another without apparently ending up with feet like pincushions.

SAVANNAH

Unlike spiny forest, 'savannah' – or, more precisely, 'savannah-woodland mosaic' – does have a number of primate specialists which are normally found nowhere else but in this particular habitat, although only where it occurs on the African continent. Here the wide, grassy plains dotted with scattered clumps of long-thorned acacia trees are the customary abode of the so-called 'savannah baboons', comprising the yellow, chacma, olive and guinea baboons. A widely spread out group of chacma or olive baboons steadily picking their way across these savannahs on their daily foraging rounds are as much a quintessential part of the African scene as the huge herds of grazing and browsing antelope and zebra. In the north-western areas of so-called 'Guinea savannah' the local Guinea baboons have to withstand a prolonged dry season accompanied by scorching temperatures from which there is little respite even by night. In a habitat where even a camera feels hot to the touch after lying in dense shade for some hours it is amazing that such well-furred animals can remain active throughout even the hottest conditions, alternately feeding and resting while an old male or two take up their posts as sentinels perched on top of large *Cubitermes* termite mounds.

The other 'savannah–woodland mosaic' specialists are the vervet monkey, patas monkey and, to a lesser extent, the chimpanzee. The patas monkey is

Patas monkeys (*Erythrocebus patas*) are found in some of the driest areas of Africa, and when water is available they drink regularly. No other primate in the world can survive in areas as dry as those inhabited by the patas.

perfectly adapted for a life spent mostly on the ground, capable of a remarkable turn of speed accompanied by an easy loping action which is unlike that of any other monkey. Patas are streetwise survivors and expert crop raiders, so the sight of a group heading out of the mangroves behind Bakau village in The Gambia, the male momentarily standing high on his hind legs to scan the surroundings, is enough to send the local women rushing for a convenient stone with which to protect the vegetable plots in which they invest so much time and effort.

An area of savannah–woodland-like formation also forms the so-called 'campo cerrado' which covers a vast area of south central Brazil. For reasons unknown there are no huge herds of ungulate to graze on the grasses African-style, and the open tracts of windswept, sun-baked grasslands do not resound to the baboon-like bark of an equivalent savannah-living primate. Where, however, a stream traverses the cerrado it nurtures a narrow but luxuriant band of gallery forest. Although this may only be 20–30 m (20–30 yds) wide it resembles in some respects a miniature serpentine linear rainforest, in which huge azure-winged *Morpho* butterflies flit lazily above the waters. Such a typical 'edge' habitat is perfect for the common marmoset, which occurs widely in these gallery forests over a huge area of Brazil. When large numbers of work-weary office workers sun themselves beside the pool in Brasilia National Park, few of them are aware of the marmosets busily snapping up insects and nibbling off gum above their hands, their fear of humans long evaporated after such regular invasions by weekend visitors.

DESERT AND SEMI-DESERT

One habitat which is virtually devoid of primates, although it covers enormous tracts in the tropics and sub-tropics, is the desert or semi-desert. With few or no trees, highly seasonal and unreliable rainfall, and little cover from the baking sun which forces most resident mammals to be nocturnal, it is perhaps not surprising that in general deserts are just too inhospitable for primates. Bonnet macaques inhabit several semi-desert areas in India but are often partly provisioned by humans, while in southern Africa the periphery of the great Karoo Desert provides a demanding and marginal living for hardy bands of chacma baboons and vervet monkeys. Perhaps the nearest that any primate comes to conquering the desert habitat are the bands of patas monkeys which survive in the rocky country in Air, which lies within the sprawling bounds of the Sahara Desert. Hamadryas baboons, too, are often called 'desert baboons', inhabiting dry, rocky country in the Horn of Africa and in southern Arabia, where they glean a spartan living from grass, leaves, fruit and the occasional insect. The broad tracts of the Kalahari, the glittering quartz and rusty sandstone deserts of Namaqualand and the ancient sand dunes of the Namib Desert on the other hand do not echo to the bark of either species. Nor do the sprawling and species-rich semi-desert areas of Mexico, the vast dry zones of the Andes or the fascinating Argentinian Monte region support even a single primate species.

TEMPERATE WOODLAND

Although most primates inhabit the tropical regions, a few species are only found within the northern temperate zones or within a temperate climatic enclave such as a mountain situated within the tropics. Temperate forests may be either deciduous, with the trees all losing their leaves during the prolonged winter, or evergreen, with the leaves, usually harder and needle-like, being kept throughout the year. Unlike in tropical rainforests, ripe fruit is not available on the trees more or less throughout the year; but it may be present on the ground (in the form of fallen acorns, for instance) or in an unripe form on the tree (such as green immature fir cones) lasting through the winter.

In the Himalayas mixed forests of both types support populations of langurs and macaques belonging to wide-ranging species capable of thriving from torrid sea-level lowlands to mountain forests which are snow-covered for several months each year. It is worth looking in detail at some of these highly adaptable species which can cope with such extremes not only of temperature but also of food supply.

For a Hanuman langur, the art of making a living on a mountainside by scraping away the snow to expose edible roots and tubers or by stripping the stringy and rather indigestible bark off young trees is a very onerous way of eking out the winter months, especially when compared with the relatively warm and easy life-style enjoyed by its relatives far down below in the monsoon forests. This incredibly adaptable langur in fact thrives from around sea level to 4267 m (13,147 ft) or so, and in at least one Indian site the animals remain above 3200 m (9850 ft) even during heavy winter snows. The olive baboon is found from the low, hot savannahs up to the cool slopes of the African mountains at 4500 m (13,860 ft), where frosts occur even in summer. The owl-faced monkey ranges up to 4600 m (14,170 ft) in the mountains of Central Africa. The highly successful vervet monkey is found from sea level to 4500 m (13,860 ft), the same height reached by the beautiful black and white colobus or guereza which is found in habitats as diverse as lowland tropical rainforest, gallery forest in semi-desert areas and highland coniferous rainforests.

In Asia the Japanese macaque lives high in the mountains and may actually migrate to higher elevations when snow mantles the ground and the temperature drops below minus 15°C. This species also lives in subtropical lowland forests, where up to 80 per cent of its diet may consist of fruit. Yet in the inhospitable habitat of the mountains on Honshu island the monkeys have to make do with a much poorer, more fibrous diet of bark, buds, leaves and whatever roots they can dig out when not covered by snow. Their only obvious adaptation to such extreme conditions is to grow a dense winter coat in the autumn, a parallel with rhesus macaques from the Himalayas which have a distinctly shaggy appearance compared with their lowland cousins.

The mountains of Asia are also home to some specialized monkeys which are superbly fitted for the high life in the forests of fir, bamboo and rhododendrons which clothe the mountainsides, for two of the three snub-nosed monkeys of western China are restricted to mountain forests. The golden snub-

The barbary macaque (*Macaca sylvanus*) is at home in the harsh environment of the Atlas Mountains of Morocco. A more comfortable living is enjoyed by the transplanted colony in Gibraltar. With such easy living they become rather fat, as seen here.

nosed monkey is found from 1500 to 3400 m (4620–10,500 ft), while the Yunnan snub-nosed monkey (*Rhinopithecus bieti*) does not seem much inclined to descend below the 3000 m (9240 ft) mark, even during the severest of winter weather; in fact at such a time the upper reaches of the forest belt between 3900 m and 4100 m (12,000 and 12,700 ft) seem to be the preferred zone for foraging. It thus seems to qualify as the highest resident primate, never coming down below 3900 m (12,000 ft) except during the spring and early summer to take advantage of the tender fresh growth of young shoots that are only available at lower levels. At higher altitudes the flower buds of the bushy mountain azaleas seem to be greatly savoured, while during the winter the Yunnan snub-nosed monkey has to eat whatever is available. During heavy snowfalls the only easily accessible food may be the long slender pendant streamers of *Usnea longissima* lichens which festoon the bare branches of the trees in these humid upland forests.

A similarly leathery and nutritionally poor diet may also fall to the lot of the Barbary macaque in the Atlas Mountains of Morocco. Here the favoured habitat is mixed forests of cedar and evergreen oak, where this hardy macaque is well able to take advantage of what little subsistence its snow-carpeted environment can offer during the winter. In bad winters this boils down to the tough, bitter needles of the cedars, supplemented by the soft layer beneath their bark which is stripped off by the teeth – virtually the only reachable provender when more desirable supplies of acorns and roots lie buried beneath the snow.

At 3800 m (11,700 ft) the Simien Mountains in Ethiopia can be a pretty harsh place to live as well, and among the giant lobelias and St John's worts which contribute much to the bizarre landscape at this altitude lives a hardy band of mammals including Simien foxes, the extremely rare Walia ibex and large numbers of gelada baboons. The gelada is confined to this highland plateau where the only abundant resource over large areas, grass, provides almost the sole source of forage; every night the bands of geladas flock back to their favourite vertical sleeping cliffs after a hard day's grass pulling.

Chapter 8
Enemies and Defence

Humans, the major enemy of primates, are dealt with more fully in Chapter 9. That apart, the degree to which primates suffer from the attentions of predators seems to depend to a large extent on their particular habitat. The dense nature of the canopy in a rainforest makes it difficult for an eagle to grab a feeding monkey from a branch, so despite the thousands of hours now spent by observers in studying forest primates there are still remarkably few actual observations of attacks by any kind of predators. The open savannahs of Africa, however, offer better prospects for a wide range of carnivores to make a successful attack on primates such as baboons and vervet monkeys, which spend a lot of time on or near the ground.

It seems likely that in many populations of forest primates predation pressure is so low as to be virtually insignificant. For vervet monkeys in the open savannahs of Amboseli in Kenya, however, life is much riskier and less predictable; it has been calculated that perhaps as many as seven out of ten vervets in Amboseli meet a premature end from attack by a leopard, eagle or python. Savannah baboons such as the anubis, on the other hand, seem to suffer less heavily but this does vary from one locality to another; at Amboseli it is presumed that up to 25 per cent are lost to predators. Generally, though, baboons are well able to look after themselves when up against any of the smaller predators. Even a single male baboon can easily see off a marauding jackal. Foolish indeed would be the jackal which stood its ground when confronted by such a yawning mouthful of stiletto-like canines; most jackals just aren't that suicidal.

The larger cats such as leopards are, however, more than a match for even the biggest and most fearsomely fanged of males, with the consequence that they are regarded as the baboons' chief enemy. But the leopard's success rate may depend very much on the bravery or otherwise of the group attacked. In some groups the big males will harass the attacking cat until it retreats; in others the largest males may well be the first to make themselves scarce, leaving their females and offspring to take their chances. The savannah baboons' defensive early warning system is also well suited to the timely detection of an incoming enemy; the sentries are usually one or more big old dominant males (alpha males), who sit at the top of a termite mound or dead tree surveying the surroundings for intruders.

Such sentinels are found in monkeys as well, and are normally also reproductive males. As time spent loafing around keeping a lookout for predators means the sacrifice of an equal amount of active feeding time, it is not perhaps surprising that it should be the dominant males who usually perform this

This female anubis baboon's (*Papio anubis*) chances of surviving an attack by a leopard may depend on the degree of valour shown by the males in her group. Some will put the big cat to flight; others will cut and run, leaving her to take her chances.

irksome task. After all, it is their offspring who may reap the benefit if an attack by a predator is quickly detected and evasive action taken; kin selection makes the rules in this instance.

In the tufted and white-fronted capuchins males spend more time than females at the periphery of the group, and less time in feeding; and they respond far more quickly and aggressively to potential threats. Sentinel behaviour has also been noted in the red-bellied lemur. This is particularly interesting because this species has a twenty-four-hour (cathemeral) routine which requires vigilance by both day and night. The group normally consists of just a single family unit – mother, father and offspring – and one of the adults is often posted on watch nearby as the other two busy themselves in feeding or enjoy a quiet snooze. The lookout executes its duties conscientiously, scanning its surroundings and moving to a fresh branch with a different viewpoint every five to ten minutes. Upon detecting a bird of prey the sentinel gives a succession of low grunts, causing the other family members either to head for cover under thick foliage or freeze for up to fifteen minutes. Nocturnal feeding sessions on the flowers of emergent eucalyptus trees borne on exposed branches also

The red-bellied lemur (*Eulemur rubriventer*) has a twenty-four-hour or cathemeral cycle of activity. A sentinel is often present while the rest of the family group feeds, probably keeping a lookout for the lemur's main enemy, the fossa (*Cryptoprocta ferox*).

demands the use of a lookout, probably keeping watch for an approaching fossa (*Cryptoprocta ferox*), a large civet-like viverrid which is an accomplished tree climber. This probably makes it the main predator upon lemurs in a variety of forest types in Madagascar.

The subject of alarm calls is a fascinating one. For example, they may not necessarily sound the same in different subspecies of a single animal. Thus in southern Madagascar Verreaux's sifaka utters its characteristic namesake 'shee-fak' call when confronted by a possible threat on the ground. Yet Coquerel's sifaka further north gives a completely different 'unsh-kiss' in the same situation. The call given by the quite newly discovered Tattersall's sifaka is almost indistinguishable from that of Verreaux's sifaka – yet the two species are separated by around 925 km (575 miles).

Response calls to potential threats have been most closely investigated in savannah-living vervet monkeys in Amboseli, where a definite 'language' seems to have evolved in which highly specific calls denote the three main types of predator and elicit the appropriate defensive response. Thus the low-pitched staccato grunting call which denotes an incoming eagle sends any monkey within earshot diving out of the trees and into thick cover. The difficult-to-describe 'leopard' call actuates the reverse response, sending any ground-feeding vervets scampering up the nearest tree and on to the slimmest, nethermost branch which will comfortably support a monkey but not a leopard. The high-pitched 'chutter' which sounds the 'snake' alarm invokes less immediate panic and more attention, inspiring any monkey on the ground to have a good look around, perhaps standing up on its hind legs for a better view. Snakes move less fast than eagles or leopards but may be harder to spot, so staying put and weighing up the situation pays off better here.

Good though this system undoubtedly is, the vervet does not depend solely on it, for there are other eyes which may also be alert to danger and other 'foreign' calls may be just as efficient at announcing an unwelcome arrival. Thus vervets also pay attention to the warning alarms of the superb starlings

Vervet monkeys have a sophisticated 'language' denoting various categories of enemy. The 'leopard' call sends any members of the group which are on the ground scampering into the nearest tree.

(*Spreo superbus*) which are so common in the same habitat. Superb starlings, too, have different sounds for different predators, a logical development when the response appropriate to an air-raid by a swooping hawk – sitting tight on the ground – would hardly be suitable when faced by a terrestrial hunter such as a civet or snake, and vice versa. So the starlings employ calls which differentiate between aerial and terrestrial predators, and the vervets tune in and decipher them just as effectively as they use their own 'monkeyspeak'.

Vervet monkeys are by no means the only primates with specific 'predator alarms'. They are probably present in all species, but have been studied in detail only in a few. Even prosimians such as the ring-tailed lemur have evolved quite a complex predator-avoidance 'language'. At its simplest this consists of a series of low clicks uttered when any kind of disturbance is noticed on the ground, without there being any clear idea about the nature of the threat – a kind of play-safe early-warning alert. Upon hearing this, any lemurs on the ground will seek safety in the nearest bush or tree before peering closely around to try to discern the threat. If this turns out to be something potentially genuine, such as a dog, fossa or strange human, then the continuing click-click utterances escalate into a loud yap; the latter will also be the first response to an immediate sighting of such a threat and has an electrifying effect, instantly sending every lemur within earshot high-tailing it for safety into the trees.

A glimpse of a Madagascar harrier hawk (*Polyboroides radiatus*) or of a

The alarm calls of the superb starling (*Spreo superbus*) are also deciphered and responded to by vervet monkeys.

Madagascar buzzard (*Buteo brachypterus*), elicits a completely different response. Both of these fairly large birds of prey have the habit of spending some considerable time sitting motionless and almost invisible in low trees or bushes. They are presumably biding their time in the hope that a likely meal (such as an unwary juvenile ring-tail) will come along. When it does, all they need to do is drop in unexpectedly on their unsuspecting victim. Ring-tailed lemurs spend a large proportion of the day on the ground; when they look up and spot one of these quietly perching raptors the response, strangely enough, is not to flee in consternation but to home in on the enemy's tree while uttering a series of 'chirps' and 'moans'. Some of the more courageous members of the group may even escalate the general level of intimidation by climbing up near the bird. When the foe eventually gets the message and flies off, the lemurs respond with a piercing blast of shrill 'shrieks'. These apparently form the standard 'harrier hawk or buzzard on the wing' call, which is audible at distances of up to 500 m (550 yds). The sudden unexpected eruption of this amazing clamour from a point nearby also makes the heart jump, a fact to which the author can attest. In marked contrast, a large flock of yellow-billed kites (*Milvus migrans*) flying overhead or even landing in a tree already occupied by ring-tails rather surprisingly elicits no vocal response whatever; the lemurs merely steal quietly away after first moving to the centre of the tree.

Although this warning system is generally trustworthy, it does seem that, as in most things, practice makes perfect. So juveniles and on occasion careless adults sometimes make mistakes and give a brief 'shriek' when it is only a harmless hook-billed vanga (*Vanga curvirostris*) or Vasa parrot (*Coracopsis vasa*) which is flying past. Juveniles obviously experience difficulty at first in distinguishing between harmless and harmful animals and may click at the bulky but innocuous radiated tortoise (*Geochelone radiata*) inoffensively if rather noisily bulldozing its way across the crisp-dried leaves on the forest floor. Juveniles also sometimes respond with the wrong call before they finally learn to get it right, perhaps yapping at a soaring harrier hawk when a shriek should have been the correct response. Adult ring-tails rarely make such mistakes except when confusion is understandable, as when a large pied crow (*Corvus albus*) with its rather hawk-like outline flies overhead and, being mistaken for a harrier hawk or buzzard, is greeted with a shriek. Such errors of judgement are also made by vervets, particularly juveniles, which may give the 'eagle alarm' to nothing more threatening than a falling leaf.

Like vervets, ring-tailed lemurs decipher the alarm calls of other species, responding to the 'hawk overhead' roar-bark of Verreaux's sifakas by striking up their own click alarm and moving closer to the safer haven in the centre of the tree, perhaps casting a few glances skywards as well. The response is two-way, for sifakas react to ring-tail shrieks by roaring. Ring-tails also react to the alarm calls of the giant coua (*Coua gigas*) and helmeted guinea fowl (*Numida meleagris*).

Crowned lemurs appear to be less able to discriminate between threats and non-threats, often exhibiting escape responses to such harmless objects as falling leaves and such relatively small and inoffensive birds as the Madagascar

Ring-tailed lemurs (*Lemur catta*) respond to the alarm calls of the giant coua (*Coua gigas*).

turtle dove (*Streptopelia picturata*). Although quite benign in nature these doves do, like most of their clan, tend to be noisy fliers, crashing clumsily about among the foliage as they take off and land. It appears to be this noise which frightens the crowned lemurs, who are seemingly unable to distinguish between large birds which are merely noisy and those which are genuinely dangerous. Crowned lemurs respond to aerial threats, real or perceived, by abandoning vulnerable positions in the exposed treetops and heading downwards as fast as possible, sometimes almost as far as the forest floor. Speed seems to be of the essence, and there is no time for the prolonged bouts of grunt-shrieking, head-craning staring and energetic tail-penduluming provoked by the detection of a marauding fossa. This particular set of signals causes any crowned lemurs near the ground to get up into the trees post-haste, for although the fossa is an excellent climber the lemurs do enjoy some advantage in the treetops. Other primates thought to employ specific signals to represent specific threats are legion, but unquestionably include the rhesus macaque, pig-tailed macaque, toque macaque and spider monkeys.

Under some circumstances a primate may take the attack to the potential adversary rather than turn tail and flee. In the Rio Blanco area in Peru a group of saddle-backed tamarins was watched mobbing a pair of quite large but non-poisonous *Corallus enydris* snakes which were innocently copulating on a liana. The whole tamarin group was involved, some individuals approaching to within 1.5 m to 2 m (1–2 yds) of the snakes. This tamarin (but this time the ssp. *illigeri*) has also been seen mobbing a related boid constricting

snake and the smaller but poisonous viper *Lechesis muta*, while even the diminutive pygmy marmoset has been seen taking on the deadly colubrid *Clelia clelia*
and giving it a hard time by some forceful and determined harassment.

During the Rio Blanco episode the group of moustached tamarins which
were usually close associates of the saddle-backed tamarins were conspicuous
by their absence. Perhaps one of the benefits for the moustached tamarin in
associating with the other species could be to cash in on its more daring and
fearless character when faced with a possible hazard. That large snakes do
indeed pose a threat to these small callitrichids is clear, for a large anaconda
(*Eunectes murinus*) has actually been seen ambushing a female moustached tamarin, despite her painstaking inspection of the fallen log which concealed the
snake before she used it as a bridge to cross a lake.

Being small, callitrichids are fair game not only for medium-sized to large
snakes but also for small cats, such as the ocelot, and raptorial birds. Attacks by
mammalian predators are probably rare and very difficult to observe, but
raids by large birds of prey happen more often and are more easily seen.
Several sizeable raptors have been watched attacking tamarins in South
American forests, eliciting strong alarm signalling which may be understood
by more than one species, with the result that saddle-backed tamarins seem to
respond to the alarm calls of their moustached relatives.

The general reaction to such alarms is to head away from exposed branches
and into the centre of the tree, or perhaps to cower beneath the handy
umbrella of a broad palm leaf. More extreme reflexes from animals badly
exposed and vulnerable on outer branches may simply be to let go and plum-
met earthwards. This may seem a somewhat drastic and potentially dangerous reaction, but is probably less perilous than it seems. Many primates seem
able to take accidental falls from remarkable heights with little apparent
harm. Coquerel's sifakas have been seen thumping on to hard, bare ground
from heights of 20 m (60 ft) or more, with no greater outward signs of distress
than to shake themselves with a startled expression before bounding uncon-
cernedly back up the nearest trunk.

Using attack as a means of defence has also been seen in other primates. In
Africa the crowned hawk eagle (*Stephanoetus coronatus*) is said to feed mainly on
monkeys. The tumultuous, panic-stricken reactions of a group of quietly feeding black and white colobus or red-tailed monkeys when they catch a glimpse
of those broad wings soaring overhead, sending the monkeys crashing for
cover among the dense inner foliage, seems to bear this out. The likelihood of
being able to benefit from the defensive capabilities of larger and more powerful species of monkeys against such formidable aerial predators may be the
main reason for the tendency of smaller-bodied monkeys to join with the
burlier, more courageous kinds to form multi-species groups.

When, for example, a crowned hawk eagle was watched mounting a sortie
against a mixed group of white-nosed guenons (*Cercopithecus nictitans*), moustached monkeys (*C. cephus*), crowned monas (*C. pogonias*) and grey-cheeked
mangabeys in Gabon it probably did not expect the welcome it received.
Instead of cowering beneath a branch, one of the white-nosed males set off

The martial eagle (*Polemaetus bellicosus*) is one of several large African eagles which attack monkeys.

crashing through the canopy in hot pursuit of the somewhat confused eagle. The chase was accompanied by a barrage of calls from most of the other monkeys in the vicinity, including a neighbouring group of black colobus. Thoroughly rattled by its unforeseen reception, the eagle blundered through the canopy before making a near crash landing in a bare tree. Loath to give in so soon, the male white-noses again briefly gave chase towards the perched eagle, but stopped short when a large mangabey male came racing athletically through the canopy towards the same goal, approaching to within 25 m (25 yds) before the unnerved raptor was put to flight.

The lesson in this little drama seems to be that the two monkeys which took the fight to their enemy belonged to the two heaviest species (6.6 kg/15 lb and 9 kg/13 lb respectively) within the mixed assemblage. On the other hand the lighter *pogonias* and *cephus* males (at 4.5 kg/10 lb and 4.1 kg/9 lb respectively) made themselves scarce along with their females and juveniles, although the rest of the *nictitans* and *pogonias* groups sat tight throughout. The smaller species, one of whom had probably been the intended victim, had therefore definitely seemed to benefit from hanging around with their bigger and more combative 'minders'.

However, even a high bodyweight does not necessarily give protection against the largest and fiercest eagles, for in Peru one of the world's most impressive raptors, the harpy eagle (*Harpia harpyja*), which weights 9 kg/20 lb, has been seen tearing apart a full-grown red howler male weighing around 7 kg/16 lb. The ability to tackle such heavyweight prey suggests that few South American primates except the burly muriqui (15 kg/33 lb) may be safe from the talons of this fearsome executioner.

With their superior intelligence, chimpanzees could perhaps be expected to have a well-developed ability to recognize and react appropriately to potential threats. Even when they are confronted by an adversary of overwhelming-

The leopard (*Panthera pardus*) often attacks ground-living primates such as baboons and vervet monkeys. However, when faced by a group of chimpanzees (*Pan troglodytes*) the large cat may be forced to flee.

ly superior weight and power, such as a leopard, they may adopt the policy of taking the action to the enemy rather than retreating in disarray. Thus a stuffed leopard was beaten unmercifully with sticks by one group of chimpanzees, while another group thoroughly demoralized a live leopard by shadowing it at a distance of 15–20 m (15–20 yds) to the accompaniment of a cacophanous volley of shrieks and visual intimidation. Even a lone female chimpanzee has been known to put a leopard to flight by pursuing it down a tree.

The acme of blatantly offensive behaviour towards a leopard, however, has been observed in the Mahale Mountains in Tanzania, where a male chimpanzee actually ventured alone into a cave to snatch a leopard club from under its mother's snarling nose; and this in the supposed security of her den. The mewling cub was the subject of a great deal of curiosity within the chimpanzee group before general mishandling led to its demise. If a single chimpanzee can seize a female leopard's most prized possession, her cub, under such adverse circumstances, it is not unreasonable to suppose that the bushbuck carcass seen in the custody of a group of Mahale chimps may have been purloined from an unhappy leopard, persuaded reluctantly to yield its hard-won kill in the face of an unacceptable level of group intimidation.

Chapter 9
Primates and Humans

To say that the primates have had and are still having a hard time from their fellow primates, humans, is something of an understatement, for roughly one-third of all primate species or subspecies are in danger of extinction as a direct result of human activities. Our sins against them include habitat destruction; logging; killing for food or, because a species is considered to be a pest, killing to produce souvenirs for tourists; killing for trophies to show prowess as a hunter; killing to collect adornments; and taking live animals for the pet trade and medical and other forms of research. All of these are making some impact upon primate numbers in the wild.

HABITAT DESTRUCTION AND LOGGING
The destruction of their habitat is almost certainly the main contributor to the reduction of primate numbers. Most primates are forest dwellers, and with the uncontrolled increase in the population of countries in south-east Asia, Africa and South America more and more of their habitat is being clear felled for agriculture. This is perhaps understandable, since people have to eat, but what is not acceptable is that a large amount of forest clearance has been perpetrated to provide cattle grazing in order to produce beef for an already overfed developed world and, even more unforgivably, to extract wood for making paper and cardboard. This had been taking place at the same time as an enormous mountain of unwanted beef has collected in coldstores in Europe. Some adaptable species of monkey in particular have been able to survive this destruction of their forest habitat by changing their habits. This has involved making use of the crops planted in the cleared areas as a source of food and later living in the secondary forest, which springs up when the soil becomes exhausted for agriculture and man moves on to clear the next area of primary forest.

The effects of logging are not always as devastating to primate populations as might be expected. It depends upon the percentage of trees taken out, the type of tree taken out, the feeding habits of the native primates and the degree to which logging opens up the forest to bring other pressures to bear on them.

In South America bearded sakis are unable to survive in heavily logged forest due to removal of their food trees, which are few and far between. Sakis, on the other hand, are less affected because they are less selective in their feeding habits. Most of the Callitrichidae seem relatively unaffected by the felling of large trees and indeed often benefit from the more open secondary forest which develops as a result, presumably because this brings in the insects which form part of their diet. Even rare species like the lion tamarin are able to hang on in highly degraded forest and then recolonize the developing secondary forest. Where logging is more selective and fewer trees are taken out, this results in less physical damage to the forest and thus it has less effect upon the native primates.

The picture in south-east Asia and Africa is similar to that in South America: those species which are most adaptable are able to hang on or even benefit from logging, while primates such as the orang-utan, so much molested by man in other ways, are badly affected by timber extraction activities. In all cases, however, logging does introduce two adverse effects on the primates as a result of the roads that are cut into the forest to enable the logs to be removed. These roads allow easy access by hunters and collectors, and it is this which often has a greater effect on primate numbers than does the logging itself.

KILLING FOR FOOD OR AS PESTS

Wherever primates are found, a proportion of them are shot to provide meat for human consumption; this practice extends even to the large apes. Chimps regularly end up in the cooking pot and even the recently discovered population of gorillas in Nigeria are suffering a similar fate – one in twenty of them are killed each year for food. Despite the reader's possible feelings of revulsion at such acts, where people are short of meat a primate is as good a source as any other mammal. In those few remaining groups of Indians who still hunt by the bow and arrow and the blowpipe, hunting makes relatively small inroads on primate populations; it is the rifle and the shotgun which do all of the damage. Earlier it was mentioned that a number of highly adapatable monkeys have taken to raiding crops where they have encroached and depleted their habitats. These monkeys often thrive in such conditions to the extent of becoming pests, and as a result they are shot and trapped by the local population.

One of the main factors responsible for the reduction in numbers of the macaques *Macaca nigra* and *M. nigrescens* on the island of Sulawesi is hunting and trapping. Both species are in fact protected by Indonesian law, but in spite of this they are still shot for food and trapped for the pet trade; in parts of the island pet macaques are a common sight. What is even more worrying is that these activities take place within the island's official reserve areas, mainly due to the reserves' inability to afford the manpower required to patrol them. Where agriculture on cleared forest has encroached on the macaques' habitat they have responded by crop raiding, which has in turn labelled them as a pest – for which reason they are also shot.

Most of the macaques are in fact hunted, and as a result they are now much less common than they once were. Hunting, for example, has been an important factor in the extinction of the most northerly population of rhesus macaques (*M. mulatta*) from Xinglung county in northern China.

Hunting in areas in Africa which were formerly inaccessible but which have now been opened up for forestry is having serious effects upon primates which were once thought to be relatively safe. The drill is endangered as a result of

Numbers of the handsome De Brazza's monkey (*Cercopithecus neglectus*) are being seriously depleted in the wild by hunting in their native Africa. Photographed at Twycross Zoo.

forest clearance for agriculture and selective logging in Cameroun. With reduction in its forest habitat the drill has taken to crop raiding, so that it is now shot both as a pest and for its tasty meat. Its natural habit of collecting in large, noisy aggregations makes it easy to find, and large numbers may be shot at one go. Other primates such as various mangabeys, both the black and white and the red colobus, Diana monkeys and even prosimians such as galagos and pottos are all suffering similar fates at the hands of the hunters. In Kenya the beautiful De Brazza monkey (*Cercopithecus neglectus*) is in danger of extinction, and its numbers are being drastically reduced by hunting in other parts of its patchy distribution in riverine forests across the central part of Africa.

179

In South America, survival of larger monkeys in areas which have been opened up for selective logging again depends as much upon consequentially introduced hunting pressure as it does on the loss of feeding trees. A further factor in this part of the world, the building of roads through the forest for mineral extraction, has also increased hunting pressure on the primate population – unfortunately in many cases for sport rather than for food.

The degree to which primates are hunted in any one area often relates to religion or local taboos. They are much less likely, for example, to be hunted in Moslem countries, although this is not always the rule. In Madagascar the hunting of any one species of lemur may depend upon the local 'fady', which dictates whether or not it is protected for one reason or another in relation to local folklore. We can be thankful that the newly discovered and rare *Propithecus tattersalli* appears to be protected by such a 'fady', for it is apparently not hunted.

KILLING PRIMATES FOR OTHER REASONS

Although it is reasonable to excuse the killing of primates by poor people as a source of food or to protect crops, killing them for other reasons is perhaps less excusable. In the past the white man has gone all over the world and fortified his macho image by shooting everything in sight, including primates, to hang on the walls of his trophy room. Even today – for a mercifully small proportion of the population – the killing of some rare creature still satisfies a peculiarly misguided need. In some areas of the world primates are still killed by some tribes to provide adornments for human apparel.

In recent years baboons and chimps have been shot in some numbers to satisfy a demand by certain unprincipled tourists for mementoes of their visit to Africa. These sick individuals are prepared to pay up to $700 for a carved human torso to which has been attached the skull of a chimpanzee; somewhat lower prices are paid if the skull is that of a baboon or other species of monkey. This trade was originally satisfied by the skulls obtained when female chimps and baboons where shot in order to obtain their babies for the pet trade and for scientific and pharmaceutical research, but now, with tourists prepared to pay such high prices for these 'curios', the primates are being increasingly slaughtered for their skulls alone. Clearly the onus is upon us so-called 'civilized' Westerners to discourage this killing by not buying these gory souvenirs. One bizarre aspect of this trade is that chimp skulls are regularly airmailed to Europe and the United States under the guise of wood carvings, and up to now their entry has never been questioned by the Customs authorities – a most peculiar state of affairs.

THE TAKING OF LIVING PRIMATES

Living primates are taken from the wild for one of two basic reasons: to satisfy the pet trade, and to provide animals for scientific and pharmaceutical research. In the past, the keeping of primates as pets by the natives of a

The cotton-top tamarin (*Saguinus oedipus*) from Colombia is one of a number of attractive primates which are being depleted in the wild by being collected for the pet trade. This individual is one of a thriving colony of this species at Twycross Zoo, where six young were successfully reared in 1990.

particular region has had a minimal effect upon primate populations, since both lived in balance with each other. But with the rapid increase in human populations in most of the Third World this demand for pets has got out of hand. If we add to this the number of primates that are required to satisfy the so-called 'developed' world, then the situation becomes untenable. At particular risk are animals such as the orang-utan and chimpanzee, for in the West these are fashionable pets for a small number of well-off members of society who think of themselves as a cut above the rest of the population. After all, what does it matter about what happens to the rest of the world and its creatures as long as they have what they want?

Taking the orang-utan as an example, it is now known that over recent years hundreds of wild-caught babies are bought and sold around the world each year, despite the fact that many countries are members of CITES (Convention on International Trade in Endangered Species) and such trade is now illegal. The poor native hunter at the beginning of this scandal is perhaps to be pitied, for the $200 or so that he receives for catching the baby, usually by first killing its mother, is to him a small fortune. From then on, however, greed rears its ugly head, for the middleman to whom the hunter sells it charges the eventual exporter around $1000 for the baby. He in his turn can expect about $5000 from the overseas importer, who can then sell the baby orang for anything between $20,000 and $30,000 to wealthy Asians and Europeans and up to $50,000 in the United States, where the law has a tighter grip on such illegal imports and accordingly the risk is greater.

Orangs are not only in demand as pets: they are also bought for zoos, where they are great attractions to visitors, and for use in circuses and theatrical acts and for films and advertising. Part of their attraction lies in the fact that they are highly intelligent creatures and can easily be trained to perform the tricks that satisfy an unfortunately large proportion of the populations of most nations. One country which does not conform to CITES is Taiwan, and in the capital, Taipei, it is said that there are more orangs per unit area than there are in their native Borneo and Sumatra. Here they are used as tourist attractions in discos and are kept widely as pets, at least while they are still small. Once they become large and unmanageable they are turned out on to the streets to fend for themselves. Most of these end up in the zoo, but the country is now taking steps to attempt to bring the situation under control and a number of orangs have been returned to their native forests. Sadly, the numbers returned are only a small proportion of those that are still being taken out.

Over the years many thousands of primates, especially species such as the rhesus macaque, have been taken from the wild for use in medical and scientific research. To a lesser extent this still goes on, though now large numbers are bred in captivity. Whether it is ethical for us humans to place ourselves in an elevated position above our fellow primates and put them to such uses is a matter of individual conscience. If their careful use reduces human suffering, then perhaps the price they have to pay is worth it, as long as we repay them by ensuring their survival in the wild. Their use for research into the safety of non-essential products such as cosmetics, however, has to be totally immoral.

PRIMATE CONSERVATION

So far in this chapter only the negative side of man's relationships with his fellow primates has been discussed – but what of the positive side, if indeed it exists? Insofar as habitat clearance and hunting is concerned, there is little that can really be done without tackling the root cause of the problem – shortage of food and overpopulation. These can only be cured by education and aid from the developed world, and by the time this has been provided it is fairly certain that a number of primates will have become extinct in the wild. If this is the case, what steps are being taken to ensure that these highly endangered species at least survive in sufficient numbers in captivity for them to be reinstated in the wild, should the opportunity ever arise again in the future? The answer is that in a number of zoos and specialist sanctuaries around the world many, though by no means all, primates are now being successfully bred in captivity.

The golden lion tamarin is in danger of extinction in the wild for two reasons: its habitat on the Atlantic coast of Brazil is fast disappearing, and large numbers of this beautiful little animal have in the past been taken for the pet trade. It was not until the 1960s that attempts were first made to halt the export of golden lions, then running at some hundreds per year, and to begin a captive breeding programme for them, for although they had been known since Brazil was first discovered by the Europeans little was known of their lifestyle in the wild. In 1972 the Smithsonian Institution's National Zoological Park in Washington had one of the world's largest captive groups of lion tamarins. In that year a conference of all interested parties was held there, during which ways in which the future of the monkey could be assured were discussed and concrete plans were produced for reintroduction of captive-bred individuals to the wild. This was followed up by the setting up of a studbook, a most important introduction for it would ensure that as often as possible outbreeding would be carried out in the captive groups.

In 1974 the Biological Reserve of Pogo das Antas was set up in Brazil as a site for the release of lion tamarins bred in captivity. Ten years later, fifteen individuals from various breeding groups in the United States were sent to Brazil for release in the reserve. Since then the number of zoos with breeding lion tamarins has increased and includes a number in the British Isles. One notable example is Twycross Zoo in Leicestershire, which boasts in its collection a total of forty-four different primate species, many of which, including the golden lion tamarin, breed successfully. On a recent visit to Twycross the authors were able to enjoy watching a pair of these attractive little animals along with their single offspring.

The success of the primate breeding programme at Twycross can be appreciated from their animal stock lists for 1990. During that year twenty-five species of primates produced offspring, though sadly some did not survive. The greatest success was achieved in the ruffed lemurs, silvery marmosets, black howler monkeys, black and white colobus, squirrel monkey, chimpanzee and, most notably, the golden lion tamarins, who produced four young.

Some centres, for historical reasons, specialize in a single species: the Monkey Sactuary near Looe in Cornwall, England, is a good example. This sanctuary breeds woolly monkeys with the aim of reintroducing as many as possible into the wild in protected areas of forest. On 14 March 1991, for example, two young males, Ricky and Ivan, left a cold, misty Cornwall at the start of a journey of thousands of miles to start a new life in the Amazonian rainforest. This journey, no doubt a traumatic experience for the two monkeys, ended some thirty-six hours later in a specially built enclosure where they were met by a familiar face in the shape of Rachel Hevesi, Director of the Monkey Sanctuary. Her words now describe the initial period of acclimatization of Ricky and Ivan to their new home.

The silvery marmoset (*Callithrix argentata*) is one of a number of species of primate which have been reared successfully in captivity at Twycross Zoo.

In their first week Ricky and Ivan were obviously very tired and quite overwhelmed by their new environment, but after a few days were calling to the free-living group [not wild but rehabilitated, confiscated Brazilian monkeys] and were soon greeting them through the wire of their cage. These initial days were spent recovering and also getting to know the sounds of the forest and some of the inhabitants, like which ants bite the hardest and which are the tastiest. The two Cornish-born monkeys displayed great nonchalance on meeting other forest dwellers such as peccaries, squirrel monkeys, spider monkeys and capuchins. After ten days on their own Ricky and Ivan were introduced to three young, orphaned woolly monkeys all under eighteen months old. At first the latter were frightened of the much larger Ricky and Ivan, having been, for the past few months, nursed back to health by humans, but quickly their confidence grew and they began to explore the enclosure, studiously ignoring one another in true woolly monkey style. It in fact took a tremendous electrical storm to bring them all together into a huddle, where they comforted one another throughout the night. The following morning the group where much more friendly and close-knit, the youngsters all playing together and then curling up with Ricky and Ivan for an afternoon siesta.

Mention has already been made of the role of Washington's National Zoological Park and other American zoos in the breeding of lion tamarins, but there are of course other centres specializing in primate breeding programmes. Notable among these is the Duke University Primate Centre in N. Carolina which successfully breeds the rarer lemur species.

In returning any zoo-bred creature to the wild a period of rehabilitation is necessary in order for it to acclimatize to what it would no doubt consider to be 'unnatural conditions'. Although certain primate activities, including some of our own, are instinctive, many of them have to be learnt. Young primates, for example, learn from their mother what is good and what is not good to eat, and without this teaching a zoo-bred animal returned to the wild would be highly likely to poison itself. One of the earliest organizations involved in such a rehabilitation was the Chimpanzee Rehabilitation Project in The Gambia, a country which lost its wild chimpanzees around the turn of the century. Since 1977 there has been a ban on the capture of or trade in wild animals in The Gambia; some of the chimps which have been rehabilitated since the Project was set up were imported illegally and then confiscated.

Inevitably, it is young chimps which are caught for trade, since adults are too dangerous to handle, and as a result they are too young to have been taught by their mothers everything that they need to know for life in the African forests. The first two chimps to join the project had been reared from a very early age with humans in the United States, and the first task was to get them to mix with other chimps. These two, in the company of Janis Carter who was to carry out their rehabilitation, went first to the Abuko Nature Reserve in The Gambia to get acclimatized to life in Africa. Here, over a period of eighteen months, they were joined by a number of other chimpanzees, some of which were home-reared while others were confiscated wild individuals.

Eventually Janis and the chimpanzees moved to Baboon Island, at 486 ha (1200 acres) the largest of a group of islands in The Gambia River, which form the River Gambia National Park. In order for the rehabilitation programme to work it was essential that Janis became the odd man out, so while the chimps roamed the island at will, she spent much of her life in a cage, only emerging to teach them the skills that they would require when they became independent of her. Basically, the chimps had to be able to recognize what was edible, to build a tree nest in which to sleep and to avoid predators such as leopards and crocodiles and other dangers such as snakes. Janis was aided in her teaching by those chimps who had been taken from the wild at an age

185

when they had already attained some of this knowledge from their mothers.

Just what Janis has had to put up with in her dedication to the chimpanzees can be appreciated from her description of teaching one of them what was good to eat. The chimp Lily would not take to eating leaves, which were essential for a balanced diet. Janis therefore had to tear edible leaves off the tree herself and then eat them whilst emitting grunts of satisfaction. On swallowing the leaves, Lily would open Janis's mouth to inspect the remains. Initially, Lily accepted some prechewed leaves from Janis, but then spat them out; it took eight months of this type of instruction before Lily accepted and ate her first leaf. Since these early days more chimps have arrived on the island, while many of the earlier arrivals have become 'wild' chimps totally independent of Janis, though still able to accept her as part of the landscape. Perhaps the most important has been the birth of the island's first baby chimps.

Projects such as those outlined above rely very much upon the blessing of the government and peoples of the countries concerned, without which they are a waste of time. At the time of writing, in 1991, the mountain gorillas in Rwanda face an uncertain future, for they are pawns in a political game being played out by government and rebel forces in this country. Despite skirmishes in their home range, the gorillas have so far not suffered any known casualties. They are, of course, a tourist attraction, so by exterminating them the rebels could help to undermine the government by ruining Rwanda's shaky economy. This is one of the arguments being used, but it is hoped both sides will come to realize that the gorillas are not there just to satisfy the tourist trade but are in themselves part of Rwanda's heritage. Sadly we in the West continue to destroy much of our own heritage, so we do not set a good example.

Though much of what one reads today about primates in the wild reports reductions in their numbers, there are one or two bright spots. The most notable of these concerns the status of the rhesus macaque in India. From an estimated 2 million monkeys in 1960 numbers fell, as a result of habitat loss and agriculture, reduction in religious protection and intensive trapping, to a low of 180,000 by 1980. Since then, as a consequence of wildlife conservation programmes and the Indian government's introduction of a total ban on their export, numbers rose to around 440,000 individuals by 1985. If only this sort of story could be written for the other more endangered species.

THE FINAL ANALYSIS

In the final paragraph of their book *Spiders of the World* the authors wrote the following: 'In the final analysis, however, spiders and their insect prey are far better equipped for survival in the long term than man himself and, whatever the shape or form of man's future follies, it is certain that spiders will still be laying traps for flies long after man has finally disappeared from the earth.'

The intervening eight years have not reduced this pessimism and, as far as the primates are concerned, whatever the shape or form of future human follies it is certain that we will ensure their extinction alongside our own, unless we learn to take care of this still wonderful world in which we all have to live.

Index

Page numbers in *italic* refer to black and white illustrations.
Page numbers in **bold** refer to colour plates.

All the photographs have been supplied by Premaphotos Wildlife, apart from the following:

Aquila Photographics
M. Birkhead 82
M. C. Wilkes 162, 175, 176

Frank Lane Picture Agency
R. Austing 33
D. A. Hosking 71
F. W. Lane 68
R. Van Nostrand 27
P. Ward 156
D. Warren 165
T. W. Whittaker 77